SpringerBriefs in Earth Sciences

SpringerBriefs in Earth Sciences present concise summaries of cutting-edge research and practical applications in all research areas across earth sciences. It publishes peer-reviewed monographs under the editorial supervision of an international advisory board with the aim to publish 8 to 12 weeks after acceptance. Featuring compact volumes of 50 to 125 pages (approx. 20,000–70,000 words), the series covers a range of content from professional to academic such as:

- timely reports of state-of-the art analytical techniques
- bridges between new research results
- snapshots of hot and/or emerging topics
- literature reviews
- in-depth case studies

Briefs will be published as part of Springer's eBook collection, with millions of users worldwide. In addition, Briefs will be available for individual print and electronic purchase. Briefs are characterized by fast, global electronic dissemination, standard publishing contracts, easy-to-use manuscript preparation and formatting guidelines, and expedited production schedules.

Both solicited and unsolicited manuscripts are considered for publication in this series.

Mehdi Zeidouni

Shale Hydrocarbon Recovery

Basic Concepts and Reserve Estimation

 Springer

Mehdi Zeidouni (iD)
Department of Petroleum Engineering
Louisiana State University
Baton Rouge, LA, USA

ISSN 2191-5369   ISSN 2191-5377   (electronic)
SpringerBriefs in Earth Sciences
ISBN 978-3-031-23558-0   ISBN 978-3-031-23559-7   (eBook)
https://doi.org/10.1007/978-3-031-23559-7

This Springer imprint is published by the registered company Springer Nature Switzerland AG
The registered company address is: Gewerbestrasse 11, 6330 Cham, Switzerland

Preface

Understanding the challenges specific to shale hydrocarbon recovery and the practices to overcome these challenges is the main focus of this book. In the first chapter, technological evolutions that led to successful production from shale plays, and the implications of the shale being a source rock for its hydrocarbon recovery are covered. The characteristics differentiating shale resources and determining play quality are also presented in Chap. 1. The second chapter presents the operations of well drilling, hydraulic fracturing, and monitoring activities. Shale hydrocarbon reserve estimation methods are covered in Chaps. 3 and 4. Chapter 3 provides an overview of the available methods for reserve estimation of shale resources followed by comprehensive coverage of decline curve analysis (DCA). The covered DCA methods include Arps' (exponential, hyperbolic, and harmonic), Stretched exponential, Power law exponential, Duong, and Logistic growth. In departure from the mostly empirical rate-time DCA methods covered in Chap. 3, advanced rate-time-pressure analysis—often referred to as rate transient analysis (RTA)—methods are presented in Chap. 4. The conventional RTA methods are extended to shale wells by introducing analysis approaches of transient linear flow, transitional flow, and stimulated-reservoir-volume (SRV) flow. Chapter 4 ends with a discussion on the complications of fluid flow in shale reservoirs and the required modelling improvements.

This book stemmed from teaching a course on *Shale Reservoir Engineering and Evaluation PETE 4190* that the author annually taught at the Craft & Hawkins Department of Petroleum Engineering, Louisiana State University since Fall 2017. *PETE 4190* also included additional contents—on the design and execution of hydraulic fracturing and its effect on well productivity—which are not covered in this book. The available literature on hydraulic fracturing stimulation were used to cover those materials. The target audience of this book include petroleum engineers and non-petroleum engineers interested in understanding the basic concepts of hydrocarbon recovery from shale reservoirs.

A note that each chapter includes questions (true/false, short answer, or multiple choice) to help the reader better engage in the learning of the presented material.

Some questions are also aimed at completing the learning process and may present new information (not covered prior to the raised question). Therefore, the reader is encouraged to answer every question as it is presented. The answers to questions are given at the end of each chapter.

Baton Rouge, LA, USA   Mehdi Zeidouni

Contents

Chapter 1
Definitions, History, and Differentiating Characteristics of Shale Hydrocarbon Recovery

Abstract In this chapter, the meaning of the term "shale" as a hydrocarbon resource is described. Next, a brief history of shale hydrocarbon recovery is presented and major steps taken to unlock its significant potential are discussed. As discussed in details in this chapter, most shale reservoirs are petroleum source rocks. The implications of the source rock nature of shale reservoirs for hydrocarbon accumulation and reservoir development is covered by explaining (1) the origin of oil and gas, (2) relative amounts and distribution of source rock resources as compared to conventional reservoirs, and (3) parameters controlling the source rock quality primarily based on the form of containing organic matter and rock pore space. The major learning outcomes of this chapter are therefore: (1) Define the term "shale" in shale hydrocarbon recovery, (2) describe the technological evolutions that led to unlocking shale recovery in the US, and (3) identify the implications of the shale as a source rock for its hydrocarbons content and type, areal extent, and recovery approach.

1.1 "Shale" Definition

In today's industry, the term "shale" is typically used to describe any fine-grain sedimentary rock that require massive stimulation for economic recovery. In this terminology shale rock may be composed of clay minerals (smectite, illite, chlorite, kaolinite), quartz, feldspar, and carbonates. Mineral compositions in the shale can be quite variable. Passey et al. (2010) and Gale et al. (2014) presented the mineral composition in major shale plays in the US (including Barnett, Eagle Ford, and Marcellus Formations) according to which the clay content varies between ~10 and 70% in these plays while the rest is quartz and carbonates.

The mineralogy of these rocks is very important for hydrocarbon recovery because rocks with less clay and more quartz and carbonates are more brittle and potentially comprised of larger pores. More brittle rocks are easier to fracture in hydraulic fracturing stimulation compared to ductile/deformable rocks. Many shale reservoirs are petroleum source rocks, mainly characterized by low permeability less than 1 micro Darcy (μD). Tight reservoirs (which are not source rocks) with permeabilities below 0.1 mD are also often categorized under shale reservoirs due to their need for extensive hydraulic fracturing to achieve economic production. In short, the "shale" terms is more about the grain size and extraction technology than the rock type and mineralogy.

Question 1.1
For which of the following, rock mineralogy is more important?

(a) High permeability conventional reservoir
(b) Ultra-low permeability shale reservoir

1.2   A Brief History

After an over 30-year decline in US oil production from 1971 peak, the shale revolution has more than doubled the US oil production to new all time high in 2019 (Fig. 1.1).

Based on production data, one may think that shale production started in the first decade of twenty first century. However, shale exploration and production started much earlier. In early 1980s, Mitchel Energy invested in the Barnett shale and drilled 800 vertical wells by 2002. Through application of massive hydraulic fracturing combined with use of low-friction stimulation fluid (slick water), economic production in Barnett shale, TX was enabled. After acquiring Mitchel Energy in 2002, Devon drilled additional wells and applied multi-fracturing in horizontal wells which proved to be very efficient in producing these resources. Natural gas production from

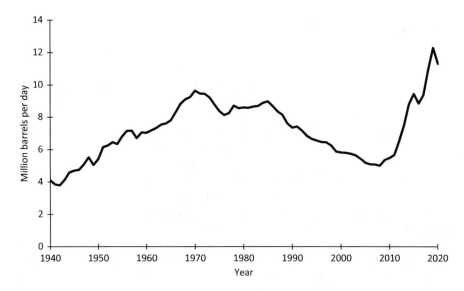

Fig. 1.1 US daily oil production since 1940 (US-EIA 2020)

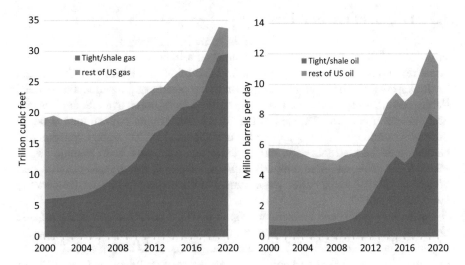

Fig. 1.2 Shale gas and oil contributions to the US production (EIA 2022)

shale started to rise in the early 2000s. However, liquid-rich shales were more difficult to develop. Innovative fracture design in the Bakken, Eagle Ford, and Permian in 2008–2012 resulted in significant production improvements of liquids from shale resources. Figure 1.2 shows how the shale contribution to total US oil and gas production evolved over time. In 2020, 70% of total US production of gas (~90 BSCF/day) came from shale while, 60% of oil (~12 million STB/day) came from shale.

Shale recovery factors are much lower than those for conventional reservoirs. The primary recovery factor of shale/tight oil is 5–8% and that for tight/shale gas is 5–20%. For comparison, the primary recovery factor of conventional gas reservoirs is 50–90%. The primary, secondary, and tertiary recovery factors of conventional oil are 12–15%, 15–20%, 4–11% respectively.

Question 1.2 (True/False)
Shale hydrocarbon recovery was enabled by the introduction of hydraulic fracturing technology.

Question 1.3 (True/False)
Liquid hydrocarbon at the surface implies that the hydrocarbon is in liquid state in the subsurface pore space.

Question 1.4 (True/False)
Liquid-rich shale recovery was enabled more recently than shale gas recovery.

1.3   Origin of Oil and Gas

In order to differentiate between the shale reservoirs as source rock, and conventional reservoirs, we need to first go through the origin of oil and gas. Carbon and Hydrogen are the most common elements in living organisms. Accumulation of dead living creatures—organic matter—occurs at the floor of ocean at a rate of a fraction of a millimeter per year. The mix of the organic matter with mineral mud at ocean floor gets buried over time. The burial is accompanied by: (1) water expulsion and hardening of the mud into rock and (2) temperature increase at an average rate of ~0.03 °C/m of burial. Assuming 0.3 mm/year of burial, the organic material gets buried at 1000 m depth after more than 3 million years while its temperature increases by ~30 °C.

At shallow depths and temperatures up to 60 °C, a small fraction of the organic matter is consumed by micro-organisms (bacteria). After eating and digesting the organic matter, the bacteria give away a natural gas known as biogenic gas. The solid residue of the organic matter known as kerogen turns into the oil and gas at higher depths. At 2–4 km burial depth and 60–120 °C temperature, the kerogen undergoes thermal cracking with increased hydrogen to carbon ratio. At temperatures near 60 °C, the kerogen turns into bitumen, and upon further heating it turns into heavy oil and then light oil as temperature gets closer to 120 °C. At 4–6 km and 120–180 °C, the kerogen turns into methane. The hydrocarbons generated between 60 and 180 °C are referred to as thermogenic oil and gas. The remaining kerogen above 180 °C will be very rich in carbon and in solid state known as coal. Note that the temperature ranges above may not be set as given because the time element needs to be taken into consideration. A given temperature maintained for a longer time period results in higher maturity than maintaining the same temperature for a shorter time period.

Question 1.5
Which of the following is the temperature range over which oil is generated at the source rock?

(a) 30–60 °C
(b) 60–120 °C
(c) 120–180 °C
(d) 180–230 °C

1.4   Source Rock Versus Reservoir Rock

The rock in which the organic matter turns into oil and gas is referred to as source rock. The source rock has a very low permeability which enables it to retain the oil and gas generated in it for millions of years. Burial of the source rock under kilometers of overlying sediments combined with the generation of lower density oil and gas causes the pressure in the source rock to significantly increase. As a

result, the source rock cracks allowing oil and gas to migrate upward to overlying permeable formations. Upward migration of oil and gas may continue all the way to the surface creating oil and gas seepages.

Alternatively, migrating oil and gas may be stopped by an impermeable layer (caprock). The oil/gas then migrates updip along the caprock until it can go longer travel further updip e.g. because of encountering crest of an anticline or a sealing fault. The resulting accumulation of oil and gas in such a structure creates the conventional reservoirs. Tens of kilometers of a migration is common for oil and gas to migrate from the source rock to the reservoir rock. From the 100% of hydrocarbon that was generated in the source rock, roughly 50% is retained in the source rock, 5% accumulates in conventional reservoirs, and 45% migrated from the source rock but not trapped. Conventional ultimate oil reserves are ~3 trillion bbls (Tbbls)—by including world oil consumption of 1.360 Tbbls to date and 1.7 Tbbls of global proved reserves. Considering recovery factor of 40% gives 7.5 Tbbls of initial oil in place in conventional reservoirs. Given the above-mentioned values of accumulations in source rock and conventional resources (50% and 5% respectively), the source rock resources would be ~75 Tbbls. Note that for simplicity, it was assumed that the ratio of oil resources to gas resources in conventional reservoirs is equal to than in shale reservoirs. Assuming 10% shale recovery factor, the shale oil reserves would equal 7.5 Tbbls. Development of shale plays has been made possible today using multi-stage hydraulic fracturing of horizontal wells. While, hydraulic fracture technology works in the US, it may not work in other parts of the world especially in arid or semi-arid regions with water access problems. Therefore, new technologies will be required to enable production of the huge shale hydrocarbon resources worldwide.

Question 1.6 (True/False)
The hydrocarbon resources in shale are ten times those in conventional reservoirs.

Question 1.7 (True/False)
Assuming recovery factors of 10% for shale and 33% for conventional reservoirs, the hydrocarbon reserves in shale are three times larger than those in conventional reservoirs.

Based on the above, two main types of hydrocarbon accumulations can be identified: (1) conventional reservoirs, and (2) unconventional shale reservoirs. Conventional reservoirs are concentrated oil and gas accumulations in high quality rocks (permeability in the range of mD–D) accommodated by a combination of natural forces including compaction, pressurizing, gravity, and structural trapping. These reservoirs are sparse, occur over relatively small areas (generally hundreds of kilometers square), and their boundaries are well constrained.

Unconventional shale reservoirs are hydrocarbon resources to be accessed in the source rocks.

With a mix of very fine grains of clay, sand, and carbonates, shales are characterized by very low permeabilities in the range of 10–1000 nD. Unlike conventional reservoirs, gravity does not play a role in distribution of fluids in these reservoirs.

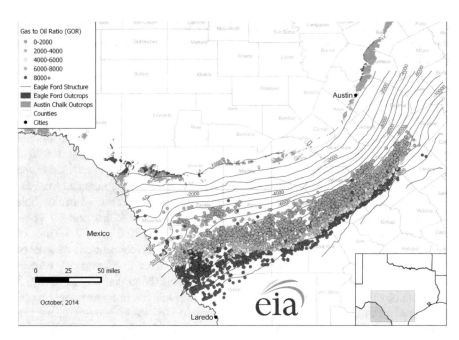

Fig. 1.3 Decreased hydrocarbon density with increased depth in the Eagle Ford shale play in Texas, USA (US-EIA 2014)

The fluid distribution is instead controlled by the depth at which the fluid was generated. As such, the fluid changes from heavy oil to light oil to condensate to gas as depth is increased (Figs. 1.3 and 1.4).

Hydrocarbon presence in shale source rocks is not constrained by the structural trapping required in conventional reservoirs. Therefore, unconventional shale reservoirs extend over vast areas. Marcellus shale (250,000 km^2) in the US is 50 times larger than the largest known conventional reservoir of Ghawar field in Saudi Arabia (5000 km^2). This implies the need for a huge number of wells to extract the shale reservoirs compared to conventional reservoirs. However, it also implies that it is less likely to drill a dry well in a shale reservoir compared to conventional reservoirs. Drilling beyond the water-oil contact or gas-oil contact in a conventional reservoir results in a dry well. In a conventional reservoir, one needs to know the boundaries of the reservoir very well and drill within those boundaries. Otherwise, the well to be drilled will be dry. In shale reservoirs, the risk of drilling dry holes is less.

Question 1.8 (True/False)
Development of a shale reservoir requires far more wells than conventional reservoirs.

Total organic carbon (TOC) is defined as the weight of organic carbon in a unit weight of rock. TOC can exceed 10 wt% for high-quality rock and is typically more than 2 wt% for shale gas reservoirs. TOC in shale reservoir can include three

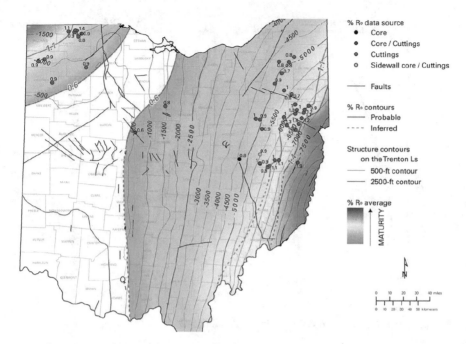

Fig. 1.4 Decreased hydrocarbon density with increased depth in the Utica shale play in Ohio, USA (Wickstrom et al. 2012)

components: kerogen, bitumen, and hydrocarbon (adsorbed or free). In conventional reservoirs, TOC is in the form of hydrocarbon only. Like conventional reservoirs, shale reservoir quality is affected by thickness, pore pressure, mineralogy, and petrophysics (permeability and porosity). However, unlike conventional reservoirs, shale reservoir quality is impacted by TOC and source rock maturity.

Burial of the source rock under kilometers of overlying sediments combined with replacement of the rock matrix by inescapable lower density oil and gas, causes significant pressure increase in the source rock. For this reason, shale reservoirs are mostly abnormally pressured. The overpressure affects the rock compaction which affects the rock porosity. In conventional reservoirs as one goes deeper, the porosity is decreased due to increased compaction. However, when encountering the source rock, the overpressure causes undercompaction. Therefore, the porosity increases with depth. This behavior is shown in Fig. 1.5 (Zhang et al. 2020). Another reason for the increased porosity is related to the secondary porosity voids created upon generation of the hydrocarbon (Charlez 2019).

Question 1.9 (True/False)
The reservoir boundaries can be well identified for source rocks.

Note that tight reservoirs are often classified under shale reservoirs which is also followed in this course. However, tight reservoirs are not source rocks. Instead, their hydrocarbon is the result of migration from source rocks to tight host sands.

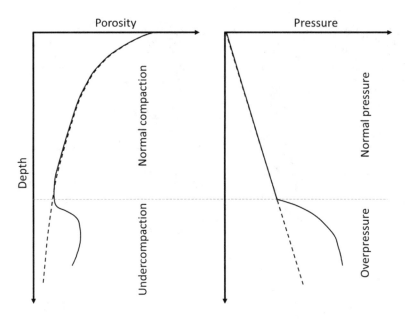

Fig. 1.5 Porosity in a mature shale play increases with depth

Question 1.10
Which of the following is not a characteristic of unconventional tight reservoirs?

(a) Permeability is below 1 mD
(b) Reservoir boundaries are diffuse
(c) Extensive stimulation is required for economic production
(d) Hydrodynamics may not control the hydrocarbon distribution

Question 1.11 (True/False)
The initial reservoir pressure located at 5000 ft depth is 5000 psia. The reservoir is abnormally pressured.

Questions Answers
1.1. b, because the mineralogy can significantly affect rock brittleness and therefore, its fracturing potential.
1.2. False. Hydraulic fracturing technology is quite old technology. The first time it was applied in the US goes back to 1947 (e.g. Zborowski 2019).
1.3. False. Both wet gas and condensate are in gas form in the reservoir but produce liquid hydrocarbon at the surface.)
1.4. True.
1.5. b.
1.6. True.
1.7. True.
1.8. True.

1.9. False. The reservoir boundaries are difficult to identify for source rocks (most shale reservoirs). This is unlike conventional reservoirs where the boundaries are set such that drilling out of those boundaries results in dry wells.

1.10. a.

1.11. True.

Chapter 2
Drilling, Completion, and Monitoring Operations

Abstract Drilling and completion of shale wells have a profound influence on well performance and associated reserves. Before introduction of massive fracturing of horizontal wells, it was not technically and economically feasible to produce most shale resources implying reserves were essentially zero. Optimizing the drilling and completion operations requires close monitoring of the operation with the aim of improving it. Different monitoring strategies must be adopted to ensure the stimulation fluids are directed to maximize the efficiency of fracturing operation. In this chapter, drilling, completions, and monitoring operations are introduced. By the end of this chapter, the reader will be able to: (1) describe how well drilling in shale/tight reservoirs differs from the conventional drilling considering well pads, well direction, and zonal isolation. (2) explain the operation of multiple fracturing (and re-fracturing) of horizontal wells and the required equipment, fluids, and proppant. And (3) discuss the methods applied to diagnose the success of hydraulic fracturing treatment using temperature, acoustic, and strain monitoring.

2.1 Drilling

Drilling process in shale reservoirs is very similar to conventional well drilling. One significant difference is the sheer magnitude of the required drilling (see Fig. 2.1). As noted in Chap. 1, shale reservoirs may be source rocks that cover vast regions. Large number of wells would then be required to drain such reservoirs. In addition, a given well can drain a much smaller area in a shale well compared to conventional well. This is mainly due to the much lower permeability in shale reservoirs which drastically decreases the mobility of fluids toward the well in shale reservoirs. It is because of lower mobility of oil compared to gas that the well spacing is generally much smaller in oil reservoirs compared to gas reservoirs. Low well spacing for shale wells is therefore necessary which often causes problems including fracture interference (fracture hits). The drainage areas for 5000 ft long lateral can vary from 40 acres (330 ft between laterals, 16 wells per mile^2) in liquid-rich shales such as Eagle Ford to 160 acres (1320 ft between laterals, 4 wells per mile^2) for gas shales e.g. Haynesville. The intense drilling and completion creates high demand for

Fig. 2.1 Dense well drilling in Wyoming (© ecoflight 2006)

materials and oilfield services and increases stress to the local infrastructure and communities.

Question 2.1 (True/False)
Shale recovery requires significantly denser drilling than conventionals.

Pad drilling, sometimes called multi-well pad drilling, is a drilling practice that allows multiple wellbores to be drilled from a single, compact piece of land known as a (drilling) pad (Fig. 2.2). Well pads are considerably cheaper than building sites for individual wells and connecting them with flow lines and roads. Well pads also significantly reduce the environmental footprint of drilling, completion, and production operations, allowing the land to be sustainably used for multiple purposes. Well pad size is dictated by on-site storage needs for equipment and fluids. Well pads commonly cover 3–6 acres.

Question 2.2 (True/False)
Drilling practice that allows multiple wellbores to be drilled from a single, compact piece of land is known as Pad drilling.

Tight sand and carbonates are commonly developed using vertical wells to access multiple productive intervals over hundreds to thousands of feet. Dense drilling may be required e.g. 64 wells per mile-square.

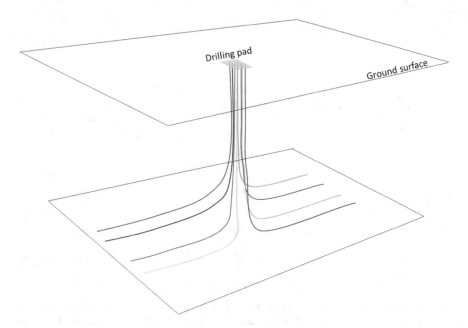

Fig. 2.2 Well pad drilling in shale reservoirs to reduce the land use and improve economics

Question 2.3 (True/False)
Horizontal well's lateral must be drilled in the direction of minimum horizontal stress so the fractures develop perpendicular to the well lateral.

One important aspect in drilling shale wells is the need to drill the well lateral in a specific direction to maximize the chance of fracture propagation in the reservoir. The stress in the subsurface can be characterized by three principal stresses perpendicular to one another. One of the stresses is the vertical stress which is simply the weight of the overburden exerted at a given depth in the subsurface. For deep reservoirs the vertical stress is generally the highest principal stress. The other two stresses will be both horizontal and perpendicular to one another. The higher stress is referred to as the maximum horizontal stress while the lower stress is the minimum horizontal stress. Therefore, the lowest stress that needs to be exceeded in order to generate a fracture is the minimum horizontal stress. Therefore, the direction of least resistance along which the fracture would propagate is perpendicular to the minimum horizontal stress. Any other direction would require higher pressure. Therefore, identifying the direction of stresses is important and the well should be drilled in the direction of the minimum horizontal stress. This enables fracture propagation perpendicular to the well lateral direction. If the well is drilled in the direction of the maximum horizontal stress, the fractures would propagate along the well (Fig. 2.3).

The success of hydraulic fracturing partly depends on the rock brittleness which is loosely defined based on Young's modulus and Poisson's ratio. Poisson's ratio (ν) is a measure of a material tendency to expand in direction perpendicular to the

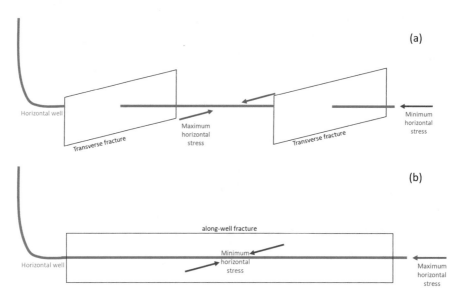

Fig. 2.3 Effect of horizontal well lateral direction on the propagated fractures (**a**) when the well lateral is perpendicular to the maximum horizontal stress, and (**b**) when the well lateral is perpendicular to the minimum horizontal stress

Fig. 2.4 Axial compression and resulting axial and transverse strain

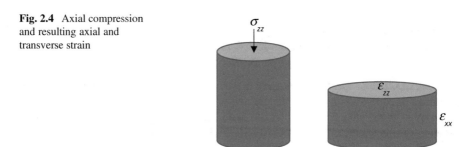

direction of compression. For compression in z direction and expansion response in x direction (Fig. 2.4), ν is defined by:

$$\nu = \frac{\varepsilon_{xx}}{\varepsilon_{zz}}$$

where ε_{xx} and ε_{zz} are the transverse and axial strains directly. ν is less than unity because the axial strain can never translate to the same transverse strain. Lower values of ν are desirable for fracturing since it implies the rock is less stretchable. For rubber ν is ~0.5.

Young's modulus (E) is a measure of rock stiffness and is defined by:

$$E = \frac{\sigma_{zz}}{\varepsilon_{zz}}$$

Higher values of E implying more stiff rock are desirable for fracturing. E and ν may be negatively or positively correlated (Burton 2016).

Shale is less brittle if it is clay-rich compared to quartz- or carbonate-rich. Tuscaloosa Marine Shale (TMS) in Louisiana is 51% clay compared to Eagle Ford where clay percentage varies between 7.6 and 19.3. TMS development has been largely hindered by the difficulty in its fracturing related to its clay content. Also, the higher Poisson's ratio in a clay-rich rock translates to higher minimum horizontal stress. This also implies that fracturing a clay-rich rock would be more difficult.

2.2 Completion (Multi-fracturing of Horizontal Wells)

2.2.1 Operation

Completions have significant effect on shale well productivity. Most common completion type is perforation of the casing along the lateral of a horizontal well and hydraulic fracturing in multiple stages. The main goal of fracturing a well is to increase the reservoir contact area. 100 stages of fracturing with 5 clusters per stage, fracture length of 500 ft, and 300 ft reservoir thickness would generate 150 million ft^2 of surface area (note that the fractures have two faces). The created fractures start healing as soon as the pumping is stopped. Proppant is needed to be pumped to keep the created fractures open.

Question 2.4
The created fracture surface area (in million ft^2) for 50 stages of fracturing with 10 clusters per stage, fracture length of 1000 ft, and 200 ft reservoir thickness is

(a) 100
(b) 200
(c) 300
(d) 400

Fracturing is conducted on stage-by-stage basis over the length of the horizontal well. There may be hundreds of stages each with an average of 3–5 perforation clusters over the length of the well which may exceed two miles. If the entire length of the well is perforated at once and fluids are injected, majority of the well length will never get fractured. This is because (1) the pumped fracturing fluids will take the path of least resistance based on the formation heterogeneities (such as natural fractures or weaknesses induced by drilling) and therefore, fractures will initiate and propagate only at selective locations along the lateral where the weaknesses

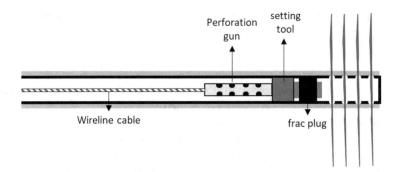

Fig. 2.5 Running wireline assembly after toe stage fracturing

exist, (2) even in absence of heterogeneities, friction losses along the lateral cause the pressure to be highest toward the heel which maximizes the chance of fracturing toward the heel compared to toe of the well. Therefore, the fracturing should be done in stages. Once a given stage is fractured, that stage needs to be isolated before the next stage can be fracture stimulated. Otherwise, the pumped fluid will continue to be taken by the created fractures.

The most common multistage hydraulic fracturing method is the plug-and-perf (PNP) method. After the well is drilled, cased, cemented, and pressure tested, the rig is moved off the location. Next, coil tubing is used to run the perforation guns downhole and well toe is perforated. An alternative to perforation is to use cementable fracturing sleeves at the toe. When pumping pressure is applied, the sleeve opens and permits entry into the formation. With fracturing sleeves at the toe, there will be no need for coil tubing. Once the connection with the formation is established, fracturing fluids are injected and fractures at the first stage are created (may take a few hours). Next, wireline assembly is lowered and pumped into the wellbore (the pumped fluids go out of the first stage fractures). The wireline assembly includes a plug behind which is the perforation gun. The plug is first set (to establish isolation) followed by second stage perforations. It should be noted that casing is generally perforated at multiple sections (clusters). Next, wireline is pulled out to the surface and rigged down. Then, pressure pumping is rigged up and stimulation fluids are pumped for the second stage fracturing. The same process is repeated in remaining stages toward the heel so that the entire length of the lateral is fractured. Once all stages are stimulated, the coil tubing is run to mill out the plugs so the well can be flowed. The fracturing stages are numbered chronologically in the order they were created so that the first stage will be at the toe and the last stage will be at the heel.

Ideally, it is desirable that all clusters in a given stage receive the same amount of fluids. Then all the stages will find equal chance of developing fractures similar to the four toe clusters shown in Fig. 2.5. However, due to pressure losses, the upstream clusters undergo higher pressure and therefore receive more pumping fluids than the downstream clusters unless the flow area is controlled with limiting the fluid entry (limited entry perforation).

One may suggest single set of perforations per stage to avoid the complications of unequal fluid distribution per cluster. While this is doable and desirable, it can significantly increase the fracturing operation time. Burton (2016) presented what this means for an example fracturing of a 9000-ft lateral with 150 ft per stage, 60 stages, and 30 ft cluster spacing (5 cluster/stage). What would be the difference between the time required to facture through these 300 points individually compared to having them grouped in stages with five clusters? Assuming 4 h overall time per stage, the individual fracturing would take [(300–60) × 4 = 960 h] = 40 days longer than five-cluster fracturing.

When fracturing a single well, the pumping crew and wireline crew cannot work in parallel and one have to await the other. However, if two or more wells are drilled which is a common practice in well pad drilling, the available wireline crew can work on another well's perforation while the pumping crew are fracturing a given well. This practice significantly improves the economics of fracturing and often referred to as "zipper fracturing".

Question 2.5 (True/False)
Fracturing is done in stages so that fractures can develop over entire length of the well.

Question 2.6 (True/False)
If hydraulic fracturing job is conducted in a single stage, fractures will be most likely created near the heel.

Question 2.7 (True/False)
Stages are labeled so that the first stage is at the heel and the last stage is at the toe.

Question 2.8 (True/False)
Limited entry perforating enables evenly distributed stimulation fluid/energy across multiple perforation clusters within a stage.

Question 2.9 (True/False)
The neighboring wells are simultaneously fractured in zipper fracturing.

Question 2.10 (True/False)
Plug and perf multistage fracturing always requires coil tubing.

Multistage hydraulic fracturing completion can be achieved by other means including ball-activated completion systems (BACS) and coil tubing activation completion system (CTACS). The reader is referred to SPE webinar by Burton (2016) for details.

In some instances, refracturing a well may be required. This may be due to a number of reasons. If the initial stage spacing is high, refracturing may be needed to fracture the inter-stage regions. In the appraisal stages, the proper stimulation treatment may not be yet known, and therefore the well may not be properly fractured. After the right stimulation strategy is identified over the development stage, the old wells may need to be refractured. This is valid for the Barnett shale where shale recovery potential was first identified. In Barnett, there has been a lot of experimental fracturing before finding what would work. Those wells are good candidates to be refractured. The fracture surface area may be lost during depletion due to stress increase, proppant loss/crushing/embedment/scaling. The well then requires refracturing to continue production.

Diverters are the most commonly used tools to conduct refracturing. Diverters are pumped to close the preexisting fractures and divert the fracturing fluid to new unfractured locations. Diverters can be disintegrating materials that will be automatically disintegrated sometime after being pumped.

Question 2.11
Which of the following does not suggest refracturing requirement?

(a) an exploratory well that has not been properly fractured earlier.
(b) a damaged well from proppant embedment, loss, or scaling.
(c) a well with initially very large spacing between stages.
(d) A well showing steep rate decline.

2.2.2   Required Equipment, Fluids, and Proppant

Hydraulic fracturing is an equipment intensive process. It requires injection of the stimulation fluids combined with proppant at variable concentrations, high rates, and high pressures. This is to be done while completing multiple wells simultaneously. Figure 2.6 shows an example well pad with hydraulic fracturing fleet on site with the following components:

1. Fluid storage is where water and other fluids (e.g. acids) are stored.
2. Chemicals indicate where different additives are stored. Fracturing fluid is composed of water, sand, and chemicals. The chemicals are required to obtain the disable fracturing fluid properties.
3. Proppant is stored in the proppant storage compartments.
4. Water, chemicals, and proppant are transported to the blender where they are mixed and prepared for injection.
5. The fracturing fluid is discharged from the blender to the pumps. The low-pressure fluid is pressurized at the pump output header (often called the "missile").
6. The required output pressure, rate, and concentrations are monitored and operated by the data van.
7. The manifold admits the pressurized fluid and directs it to the wellhead.
8. Wireline unit is to isolate previous fracture stages while introducing new perforations to allow communication between the wellbore and the formation at the current stage (see Fig. 2.5).

The large volumes of fluids and proppant are required and consequently, the equipment to inject them have dramatically increased over time. Table 2.1 shows a

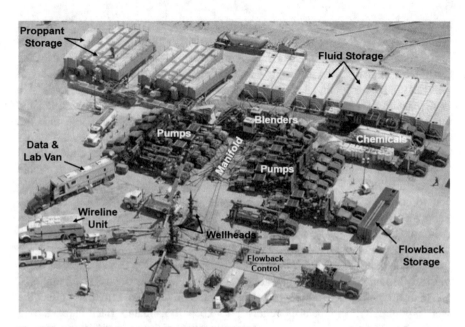

Fig. 2.6 A hydraulic fracturing site (Elwaziry 2020)

Table 2.1 Comparison of multi-fracture horizontal well completion requirements in 2012 versus 2017 for 10 basins (Weijers et al. 2019)

Parameter	Unit	2012 average	2017 average
Lateral length	ft	5580	7625
Stage count		19.3	38.6
Proppant mass	lb	3,506,284	11,891,000
Proppant mass per lateral foot	lb/ft	677	1632
Fluid volume	bbl	74,411	243,983
Fluid volume per lateral foot	bbl/ft	14.4	33.2
Average proppant concentration	PPG	1.17	1.21
Maximum rate	bpm	57.6	81.7
Maximum rate per lateral foot	bpm/ft/ stage	0.2	0.42
365-day cumulative oil	BO	61,044	108,209
365-day cumulative oil per lateral foot	BO/ft	12.2	17.7
365-day cumulative oil equivalent	BOE	91,465	159,942
365-day cumulative oil equivalent per lateral foot	BOE/ft	18.2	25.7
Well cost	Million $	7.2	5.1

comparison between 2012 and 2017. You will find videos and additional visualizations of the equipment in the assignment of this Module.

Assuming the first year recovery to be half the expected ultimate recovery (EUR) would give an EUR of 320,000 BOE per well in 2017 (Table 2.1). This is only 30% larger than the average amount of fluids pumped (~244,000 bbl) to stimulate the well. Also, when dividing the average fluid volume per well (243,983) to stage count per well (38.6), we obtain 6320 bbl/stage which is equivalent to ~1000 m^3/stage (this is a water in $10 \times 10 \times 10$ m^3 container). The main objectives of pumping the stimulation fluids are to propagate the fracture as deeply as possible into the reservoir, while transport the proppant along its length. An unpropped fracture loses its aperture soon after production begins and provides minimal conductivity. The fluid therefore must have proper viscosity and low friction pressure to enable propagating the fracture while carrying the proppant along. In addition, the fluid should provide good fluid loss control to minimize the fluid loss to the reservoir. Otherwise, the fluid will be lost to the formation rather than propagating the fracture. Additives including gels and crosslinks are used to increase the viscosity of the fluid when pumped. These additives should be able to quickly breakdown when the treatment is done. In the field "slickwater fracturing" is often used to refer to a low viscosity stimulation, while "gel fracturing" is used for high viscosity stimulation.

As mentioned above, unless the created fracture is propped, it may not affect the well productivity. Proppant is required to keep the created fracture open and conductive. Most commonly used proppants are sands with various mesh sizes. Irregular size and shape of sand proppants may reduce the conductivity. Ceramic proppants can provide more uniform size and shape and thus higher conductivity. Ceramic also offers higher strength and is more resistant to crushing. Also, adding resin helps to consolidate the proppant and prevent sand flow during production. Division of the average proppant mass required per well (12×10^6 lb) by the number of stages (38.6) gives 310,000 lb per stage. This is equivalent to about 1.5 train carload.

2.3 Monitoring and Near-Wellbore Diagnostic

Monitoring the drilling, hydraulic fracturing completion, and production operations is required to optimize the shale hydrocarbon recovery. Monitoring can address a number of unknowns especially on the cement quality, stage isolation, slurry allocation, proppant allocation, well interference (frac-hit), fracture geometry, and flow contribution per individual fractures.

The information monitored downhole using the available monitoring technologies today are mainly temperature, acoustic, and strain. Fiber optic (FO) sensing allows acquiring these data simultaneously and continuously in real-time fashion with high temporal and spatial resolutions. The data are often used jointly to answer the questions of interest.

FO sensing is based on transmission of light—with specific wavelength—through a fiber cable. A pulse of light generated by an optical source is sent down the fiber. The backscattered (refracted) light is received which carries information on

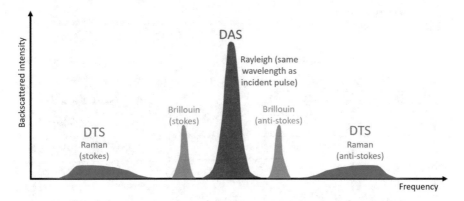

Fig. 2.7 Various backscattered lights (Rayleigh, Brillouin, and Raman) enable extract physical properties of the fiber and translating them to measurements of temperature, strain, and acoustic signals along the length of the fiber (based on Weber et al. 2021)

the properties of the fiber cable (its density and molecular vibration) along the entire cable. Note that the entire fiber cable acts as sensor; hence the cable is a distributed sensor, and the measurement is a distributed measurement. The distributed temperature, strain, and acoustic sensing are often abbreviated as DTS, DSS, and DAS, respectively.

The backscattered light in response to the light pulse comes at different frequencies. The one with the same frequency as the light pulse which has been sent down is called Rayleigh backscatter. The other two backscattered signals are referred to as Brillouin and Raman (see Fig. 2.7). These backscattered lights provide a measure of changes in the density and molecular vibration of the fiber along its entire length, which is then translated to distributed measurements of temperature, strain, and acoustic along the entire length of the fiber.

Fiber optic cables can be deployed permanently or temporarily either behind the casing, insider the casing, or attached to the tubing. Detailed information on FO sensing can be found in the extensive literature devoted to the topic e.g. Hartog (2017).

2.3.1 Expected Temperature, Strain, and Acoustic Signals

During hydraulic fracturing treatment, cooler fluids are injected and therefore temperature drop is expected. During shut down, the temperature should recover and therefore, warmback is expected. After hydraulic fracturing stimulation, and during the production period, the temperature response can be more complex. The temperature at any given location would depend on the temperature and rates upstream of that location, the geothermal gradient, the produced fluid type (liquid versus gas), among other parameters/processes.

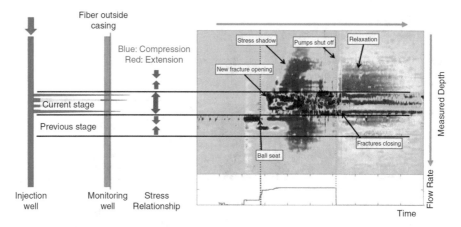

Fig. 2.8 Strain data showing extension at the fracturing stage and compression at the neighboring stages (Raterman et al. 2017)

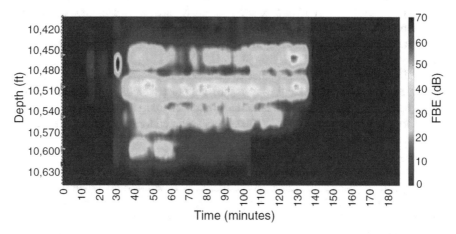

Fig. 2.9 The FBE calculated from fiber acoustic data (Pakhotina et al. 2020)

Rock expands at the location at which fracture is initiated while it contacts on the surrounding. Therefore, positive strain is expected at the fracture location and negative strain should be observed in the surrounding rock (see Fig. 2.8).

It has been established that there is a relationship between the flow rate and the sound pressure level which is obtained by transformation of raw acoustic data to frequency band energy (FBE). The waterfall plot in Fig. 2.9 shows an example FBE calculated from DAS data along a multifractured horizontal well. In all waterfall plots shown here, y-axis indicates depth and the x-axis shows time. The data are for a single stage with four perforation clusters. The FBE indicates the intensity of the flow at each of the clusters.

Below are some examples of using DTS, DSS, and DAS data to extract information relevant to recovery of hydrocarbons from shale assets.

2.3.2 Frac-Hit Identification

When fracturing a well (child well) in presence of previously fractured well (parent well), the stress shadow may cause the fractures at the child well to hit the parent well—which is often referred to as frac-hit or fracture interference in the field. Figure 2.10 shows the strain data acquired along a parent well during fracturing of a child well. The positive strain indicates expansion and negative strain implies contraction. Prior to the frac-hit, the rock expands at the parent well. After the frac-hit occurs, both expansion and contraction are observed. The expansion happens at the location intersected by the fracture while the surrounding rock experiences contraction. The rock contracts afterwards during fracture closure (Shahri et al. 2021). The strain changes have been inverted to obtain the width of intersecting fracture (e.g. Liu et al. 2022).

2.3.3 Fracture Geometry

The acoustic data can be visualized to capture the low-magnitude (from −4.0 to 0.0) seismic events induced during the fracturing. These microseismic events can hint to where the fractures developed and provide an estimate of their geometry. The microseismic events may be available from FO DAS data. However, microseismic can be alternatively captured by geophones. Figure 2.11a shows geophones in a vertical well close to the well lateral to be fractured. The resulting microseismic data are shown for a single stage in Fig. 2.11b. The region where the seismic events are occurring is often referred to as microseismic cloud which can hint to the stimulated

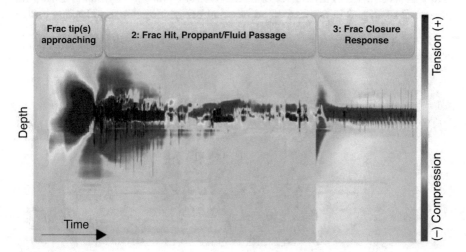

Fig. 2.10 Expansion (tension) and contraction (compression) observed prior and after fracture hit (Shahri et al. 2021)

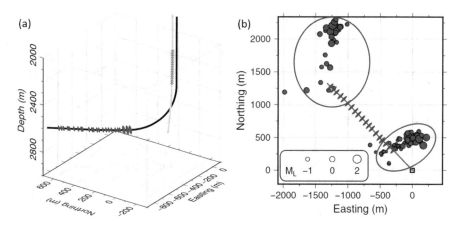

Fig. 2.11 (**a**) A multifractured horizontal well where blue and red show the stages. The geophones (purple triangles) deployed along a nearby vertical monitoring well (pink line). (**b**) The microseismic cloud corresponding to the stage shown in red (Chen et al. 2018)

reservoir volume (SRV). The overall SRV derived from the dispersed microseismic events is commonly overestimated in size and complexity. A single planar fracture may be mapped as scattered events, that is, the semitic cloud may represent more complexity than the actual fractures (Chen et al. 2018).

2.3.4   Stage Isolation

The two panels in Fig. 2.12 show DTS waterfall plots for hydraulic fracturing treatment of stage 2 in presence of an isolating bridge plug (top) and leaking bridge

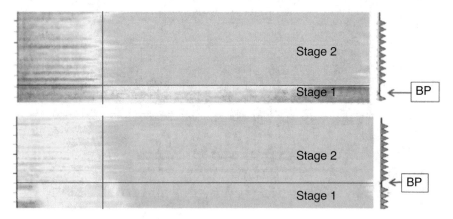

Fig. 2.12 Temperature response during hydraulic fracturing of stage 2. Black lines indicate the location of bridge plug (BP) and the red vertical lines indicate the starting time of hydraulic fracturing (rearranged and modified from Sookprasong et al. 2014)

plug (bottom). At the beginning (time $= 0$), both panels show that the temperature over the entire depth of the well is almost uniform. The warm colors show higher temperatures and cool colors show cooler temperatures. The start of pumping at stage 2 is recognizable by the quick cool down due to injection of cooler stimulation fluids (marked by the red vertical lines). The top panel shows that the temperature cooling during pumping the second stage remains within the stage and does not appear in the first stage. This indicates very good isolation between stages 1 and 2. On the contrary, for the bottom plot, cool down in stage 1 appears shortly after the start of pumping stage 2 which implies poor isolation. This is either due to cement integrity issues, improperly working plugs, or longitudinal fractures. More examples can be found in the literature e.g. Raterman et al. (2017).

2.3.5   Slurry Allocation and Fracture Initiation Points (FIPs)

While the temperature changes during the treatment is of interest for isolation, the temperature warmback just after hydraulic fracturing can be used to locate the developed fractures. Faster warmback implies that less slurry was admitted and propagated. Slower warmback then can indicate the locations where the fractures have been initiated and propagated. Information on the fracture initiation points (FIPs) can be extracted from warmback data immediately following the HF. This data gives the operator a chance to vary fracture design parameters (e.g. pumping rate and pressure) for various stages and see the effect on the number of FIPs. The stage can then be compared in terms of FIP spacing (average distance between adjacent fractures), fracture intensity (FIP/ft), and cluster efficiency (FIP/number of clusters). The first two measures are essentially the same and can be used interchangeably.

In Fig. 2.13, FIPs are shown immediately and 10 days after the HF treatment of a well. As expected, larger temperature difference between stages is visible immediately after the fracturing compared to 10 days after. Distinct temperature minima are identified as FIPs. It should be noted that better FIPs metrics cannot ensure better fracture networks unless assuming that all fractures are equally propped.

DAS data can be also used for slurry allocation as was shown earlier in Fig. 2.9. As noted earlier, the slurry allocation is not synonymous with stage/cluster contribution to production. The DAS and DTS during the production period can be used to determine the stage/cluster contribution to production.

2.3.6   Production Profiling

FO data can be used to determine the rate per individual fracturing stage or cluster.

Mao et al. (2021) presented analysis of DTS data acquired during early production (flowback) period to determine the production profile along fractured well lateral. Hveding et al. (2020) used both DAS and DTS data for the same purpose.

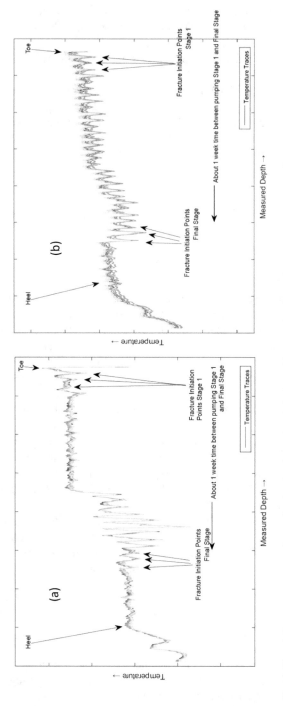

Fig. 2.13 (**a**) Temperature traces immediately after fracturing and, (**b**) at 10 days after fracturing (Natareno et al. 2019)

2.3.7 Other Types of Data

FO data has been highly underutilized and there is huge potential to improve their use to manage the shale recovery. In recent years, downhole cameras have also been deployed for downhole monitoring. It should be noted that bottomhole pressure during hydraulic fracturing treatment has been traditionally used to obtain information on the fracture growth and estimates of primary fracture parameters. Pressure combined with rates are also used during flowback and production periods to extract information on the overall fracture stimulation (which will be partly addressed in Chap. 4). However, it is very difficult to extract information from the pressure response on *individual* fractures and their contribution to production. In the monitoring data above the focus has been on extraction of information per individual fracture both during and after fracturing.

Questions Answers
2.1. True.
2.2. True.
2.3. True.
2.4. b, because $50 \times 10 \times 1000 \times 200 \times 2 = 200e6$ ft^2.
2.5. True.
2.6. True.
2.7. False.
2.8. True.
2.9. False.
2.10. False, because if the toe stage is equipped with frac sleeve, there will be no need for coil tubing.
2.11. d.

Chapter 3
Reserve Estimation Through Rate-Time Analysis

Abstract In this chapter and the next, the methods for hydrocarbon reserve estimation of shale resources are covered. In this chapter, we first review the basic definitions relevant to resources and reserves, and the available methods for their evaluation. The importance of rate-time analysis techniques (e.g. Arps' decline curve analysis) to estimate shale reserves is discussed. Next, we introduce and apply the most commonly used rate-time analysis methods in shale reserve estimation including, Arps' (exponential, hyperbolic, and harmonic), Stretched exponential, Power law exponential, Duong, and Logistic growth.

3.1 Resources and Reserves

Resources are the quantities of hydrocarbons initially in place. Reserves are the remaining hydrocarbons to be produced. The hydrocarbons must be physically and economically producible for them to be called reserves. Original reserves at the time of discovery which are going to be produced over the lifetime of a field/well are referred to as estimated ultimate recovery (EUR). The ratio of EUR to resource is referred to as recovery factor (RF). Reserves at any given time are the EUR minus the cumulative production.

For conventional reservoirs, the reserves and resources are field-scale metrics. That's because drainage area of a given well and consequently its production is affected by neighboring wells activities. Therefore, resources and reserves do not apply to a conventional *well*; they apply at *field* scale. For a developed field, and assuming stabilized rates, the resource per well (N_j for j^{th} well) can be assigned based on its corresponding rate $N_j = \frac{q_j}{q_T} N$ where N is the total resource, q_T is the total production rate, and q_j is the production rate of the jth well. However, given the very low permeability of shale resources, a given well drainage area can be considered as fully separate from the neighboring wells, unless there is significant interference e.g. due to fracture hits. For this reason, the resources and reserves are used at *well* scale for shale resources. In this book, we use the reserves/resources at well scale considering its applicability to shale wells.

M. Zeidouni, *Shale Hydrocarbon Recovery*, SpringerBriefs in Earth Sciences,
https://doi.org/10.1007/978-3-031-23559-7_3

The main methods to calculate resource and reserves include volumetric, material balance, production data analysis, and analogy.

Volumetric method is based on the static properties including the formation thickness (h), drainage area (A), initial water saturation (S_{wi}), porosity (ϕ), and initial formation volume factor (B_{oi} or B_{gi}). Except for the last parameter, all other properties can be derived from well log evaluations. For an oil reservoir, the initial in place volume i.e. the resources (N) can then be obtained by:

$$N = \frac{Ah\phi(1 - S_{wi})}{B_{oi}} \tag{3.1}$$

To obtain EUR, the recovery factor is needed. Recovery factor is not available from static data and requires knowledge of numerous variables including operational constraints, recovery mechanisms, fluid property evolution, and reservoir heterogeneities.

Unlike volumetric methods, material balance (MB) technique requires dynamic data including the average pressures and corresponding produced volumes. Fluid properties and their evolution with pressure are also required. MB can be used to estimate N (G), EUR, RF, and water influx rate. More importantly, it can be used to estimate the future production versus average pressure. The major problem with MB is that it relies on average pressure which is difficult to measure for shale reservoirs. Flowing material balance (FMB) has been introduced to overcome this limitation. In FMB, the well bottomhole pressures are used instead of average pressure. This allows application of MB to shale wells.

Production data analysis requires rate history for rate-time analysis (often referred to as decline curve analysis, DCA), and history of rate and bottom hole flowing pressure for rate-time-pressure analysis (often referred to as rate-transient analysis, RTA). For DCA, rate decline versus time and/or cumulative production are studied. For RTA, pressure-normalized rate versus time and/or cumulative production are analyzed. Therefore, DCA and RTA are applicable for reserve estimation of shale wells and will be discussed in detail herein.

Analogy is often used in shale wells based on a type well decline which represent the average decline behavior of the wells drilled in the play. Analogy can also be done in terms of recovery factor, barrels per acre, EUR, etc.

Among the above reserve estimation methods, production data analysis is widely used for reserve estimation of shale wells. MB can only be used if the average pressures are available which would require very long shut-ins. Application of the volumetric method may be difficult because of difficulty in obtaining independent estimate of static data and recovery factor.

Question 3.1
Which of the following reserve estimation methods requires the recovery factor?

(a) Volumetric
(b) Material balance
(c) Production analysis

Question 3.2
Which of the following reserve estimation methods cannot be done before drilling?

(a) Volumetric
(b) Material balance
(c) Production data analysis
(d) b and c

3.2 Hydraulic Fracturing Effect on Well Productivity and Reserves

It is important to identify how the fracturing stimulation may affect the reserves. In terms of fluid flow, fracturing (1) changes the flow regime from near-wellbore radial to linear or bilinear, and (2) eliminates skin effects by bypassing near-wellbore damage zone. The fracture geometry can be idealized assuming that the developed fracture is a bi-wing planar fracture. The fracture geometrical characteristics would then be the fracture height, width, and length. The height is often constrained by the vertical heterogeneity of the layers and is not a characteristic to be optimized. It is of prime interest to determine the fracture length and width that would maximize the well productivity. In addition, the fracture permeability (which depends on the proppant filling it) is important parameter that controls the ease of the flow in the fracture. The desirable fracture geometry can be qualitatively explained through analogy to road requirement to carry given traffic. A note that, considering the bi-wing fracture, the fracture half-length (x_f) of the fracture is often used to characterize the fracture rather than the fracture length. The fracture width is denoted by w_f.

Consider a densely populated area versus sparsely populated region as shown in Fig. 3.1. What are the desirable road length and width to help carry the traffic in these two areas? For densely populated region, faster and wider road is more important for carrying the traffic than road length. For sparsely populated region, longer road is more important to traffic flow than road width and quality. In comparison to hydraulic fracturing, the road width, road length, road quality/speed limit, and surrounding population are equivalent to fracture width, fracture length, fracture permeability, and reservoir permeability. For conventional reservoirs, k_f and w_f are more important to stimulation effectiveness than x_f. In shale, x_f is more important than k_f and w_f. These properties are grouped into an important dimensionless number known as dimensionless facture conductivity:

$$C_{fD} = \frac{k_f w_f}{k x_f} \tag{3.2}$$

C_{fD} can be thought of as the ratio of the highway's ability to carry traffic to the ability of the feeder system to supply the traffic. C_{fD} must be larger than unity, obviously.

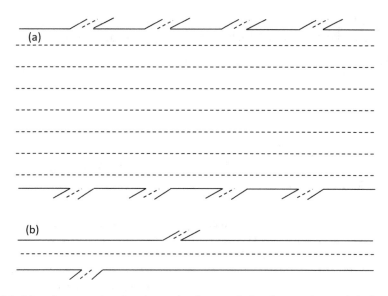

Fig. 3.1 (**a**) road geometry in a densely populated area equivalent fracture characteristics in a high-permeability reservoir, and (**b**) road geometry in a sparsely populated area equivalent fracture characteristics in a low-permeability reservoir

For moderate-permeability reservoirs (up to 50 mD for oil and 1 mD for gas), the fracture accelerates production without impacting well reserves (see Fig. 3.2). However, in low-permeability reservoirs (less than 1 mD for oil and 0.01 mD for gas), fracturing contributes both to well productivity and to the well reserves (see Fig. 3.2).

Production data analysis can be performed in terms of rate-time analysis or rate-pressure-time analysis. As noted above, the former is often referred to as decline curve analysis (DCA) and the latter is referred to as rate transient analysis (RTA). DCA is the subject of this chapter and RTA will be introduced in the next chapter. DCA is a method to estimate reserves by analyzing and extending the past decline trends of production rates. The required data are the production rates and cumulative productions versus time. DCA techniques are generally empirical although, at some instances, they can be derived from fundamental equations. For the DCA to be applicable, the well conditions (e.g. well damage, choke size, pump, artificial lift) and reservoir conditions (e.g. production mechanisms, flow regimes, flow geometries) should be stable. The well bottomhole pressure should be also assumed constant. Because of these restrictions, the early flow period during which well conditions are unstable should be excluded from DCA. The early flow period can be in the months' range for shale wells.

In this chapter, we cover the commonly used DCA techniques for shale wells which include Arps', modified Arps' (segmented), stretched exponential, power-law exponential, Duong, and logistic growth.

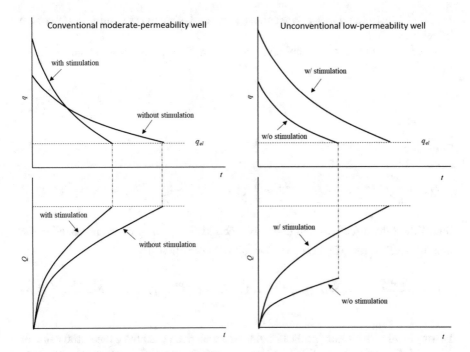

Fig. 3.2 The impact of hydraulic stimulation on the reserves and productivity of conventional and unconventional shale wells (modified from Economides et al. 2013)

3.3 Arps' DCA

In 1945, Arps presented three rate decline behaviors namely exponential, hyperbolic, and harmonic. These three decline behaviors are discussed in this section. In the next chapter, we show how the exponential decline can be derived from solving the fundamental governing equations.

3.3.1 Arps' Exponential DCA

According to Arps exponential DCA, the rate declines exponentially based on the following equation:

$$q = q_i e^{-Dt} \rightarrow ln\left(\frac{q}{q_i}\right) = -Dt \text{ or } log(q) = log(q_i) - \frac{Dt}{ln(10)} \quad (3.3)$$

where q_i is the initial production rate, q is the production rate after a given time t, D is nominal decline rate. Values of D can easily exceed 1/year for shale wells. Average

D of OPEC giant fields is ~6%/year i.e. rate by end of first year would be 94% of the initial rate.

In order to obtain the cumulative production, the rate equation should be integrated over time:

$$\int_0^t qdt = \int_0^t q_i e^{-Dt} dt \tag{3.4}$$

Therefore,

$$Q = q_i \int_0^t e^{-Dt} dt = -\frac{q_i}{D} e^{-Dt}\big|_0^t = \frac{q_i}{D}\left(1 - e^{-Dt}\right) = \frac{q_i}{D} - \frac{q_i e^{-Dt}}{D} = \frac{q_i - q}{D} \tag{3.5}$$

Sometimes, the effective decline rate is reported ($D_e = \frac{q_i - q}{q_i}$) which is related to the nominal decline rate (D) based on the following relationship:

$$D_e = \frac{q_i - q}{q_i} = 1 - \frac{q}{q_i} = 1 - e^{-Dt} \to 1 - D_e = e^{-Dt} \to D = \frac{-\ln(1 - D_e)}{t} \tag{3.6}$$

Example 3.1 The initial production rate of a well decreased from the initial value of 1200 MSCF/day to 1134 MSCF/day after 100 days of production. Calculate (a) D and D_e. (b) q and Q after 2 years. (c) well production life if the economic limit rate (q_{el}) is 2 MSCF/day.

Solution
(a)

$$\ln\left(\frac{q}{q_i}\right) = -Dt \to \ln\left(\frac{1200}{1134}\right) = -D\,(100\ \text{day}) \to D = \frac{0.0566}{100\ \text{day}}\ \frac{365\ \text{day}}{\text{year}}$$

$$= 0.206/\text{year}$$

$$D_e = \frac{1200 - 1134}{1200} = 0.055.$$

(b)

$$q = q_i e^{-Dt} \to q = 1200\ \exp\left(-0.206 \times 2\right) = 794\ \text{MSCF/day}$$

$$Q = \frac{q_i - q}{D} = \frac{(1200 - 794)\text{MSCF/day}}{0.000566/\text{day}} = 717{,}649\ \text{MSCF}$$

(c)

$$t = \frac{\ln{(q_i/q_{el})}}{D} = \frac{\ln{(1200/2)}}{0.206} = 31 \text{ years}$$

Question 3.3
What percentage of the initial rate would the rate be after one year if $D=1$/year?

In addition to the exponential decline, Arps presented the hyperbolic and Harmonic declines to be discussed next.

3.3.2 Arps' Hyperbolic and Harmonic DCA

Exponential decline implies that plotting log of rate versus time should provide a straight-line. However, this is generally not the case. Also, it has been observed that the exponential decline results in conservative estimates of EUR and well life. Hyperbolic and harmonic equations were introduced to extend the exponential decline. The hyperbolic decline was introduced based on:

$$q = q_i(1 + bD_it)^{-1/b} \tag{3.7}$$

Di is the initial decline rate and b exponent is a measure of change in decline rate. Note that the above equation reduces to the exponential decline when $b = 0$ because taking natural log of the Equation above gives:

$$\ln{\left(\frac{q}{q_i}\right)} = \frac{-1}{b} \ln{(1 + bD_it)} \tag{3.8}$$

Assuming $b = 0$ makes the RHS undefined. Using L'Hôpital's rule, and taking the derivative of the numerator and denominator with respect to b, we obtain:

$$\lim_{b \to 0} \frac{-1}{b} \ln{(1 + bD_it)} = \lim_{b \to 0} \frac{-D_it}{1 + bD_it} = -D_i \tag{3.9}$$

Therefore, $\ln{\left(\frac{q}{q_i}\right)} = -D_i$ which is in identical to the exponential decline equation.

When $b = 1$, the hyperbolic equation above reduces to harmonic decline equation. Figure 3.3a shows the rate ratio (q/q_i) for various b-values considering $D_i = 1/$year.

The b-value should be between 0 and 1 if flow is boundary dominated. b-values larger than 1 indicates transient flow. Multifractured horizontal shale wells may present b values that approach 2. b is shown to be 2 for transient linear flow and 4 for transient bilinear flow. D_i is constant and for shale wells easily exceed 1/year. If $b = 0$, the decline rate will be fixed at D_i. Increasing b-value decreases the rate of decline with time based on the following:

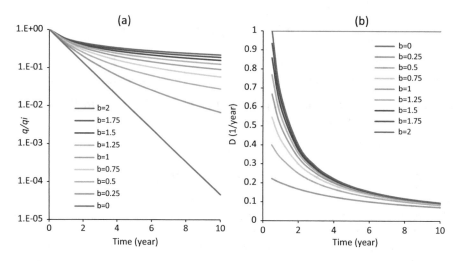

Fig. 3.3 Time-dependence of (**a**) rate ratio, and (**b**) nominal decline rate for various b-values

$$D(t) = \frac{-1}{q(t)} \frac{dq}{dt} \rightarrow D = \frac{D_i}{(1 + bD_i t)} \qquad (3.10)$$

Note that D is no longer constant. It decreases with time i.e. the rate decline will be slowing down. Figure 3.3b shows how decline rate (D) varies with time for different values of b.

The equation for cumulative production based on hyperbolic decline is given by:

$$Q = \frac{q_i^b}{(1-b)D_i} \left(q_i^{1-b} - q^{1-b} \right) \qquad (3.11)$$

For harmonic, $b = 1$ which gives (by taking the limit of the expression above for $b \rightarrow 1$):

$$Q = \frac{q_i}{D_i} \ln \left(\frac{q_i}{q} \right) \qquad (3.12)$$

Assuming hyperbolic decline over the entire production period of a shale well can results in highly unrealistic EUR and well life (Wright 2015). For instance, consider a shale well with $q_i = 7000$ MSCF/day, $q_{el} = 10$ MSCF/day, $D_i = 0.022$/day, $b = 1.4$. The EUR and well life would be:

$$EUR = \frac{q_i^b}{(1-b)D_i}\left(q_i^{1-b} - q_{el}^{1-b}\right) = \frac{7000^{1.4}}{(1-1.4)(0.022)}\left(7000^{(1-1.4)} - 10^{(1-1.4)}\right)$$

$$= 10.1 \text{ BSCF}$$

$$t = \frac{\left(\frac{q_i}{q_{el}}\right)^b - 1}{bD_i} = \frac{\left(\frac{7000}{10}\right)^{1.4} - 1}{1.4 \times 0.022} = 312{,}278 \text{ days} = 855 \text{ years!}$$

This well life is obviously unrealistic. To resolve this issue which is expected for large b values—often encountered for multifractured horizontal shale wells—switching from hyperbolic to exponential decline is required at (1) a given decline rate, or (2) a given time.

Switching from hyperbolic to exponential when a given terminal decline rate (D_{trm}) is reached requires obtaining the time at which the switch should occur which is given by:

$$t_{\text{trm}} = \frac{\frac{D_i}{D_{\text{trm}}} - 1}{bD_i} \tag{3.13}$$

An alternative approach to switching from hyperbolic to exponential decline is to do so at a specific time, t_{trm}. The corresponding switching D is then given by:

$$D_{\text{trm}} = \frac{D_i}{1 + bD_i t_{\text{trm}}} \tag{3.14}$$

Example 3.2 Considering hyperbolic decline with $q_i = 7000$ MSCF/day, $q_{el} = 10$ MSCF/day, $D_i = 0.022$/day, $b = 1.4$. Calculate EUR and well life, (a) assuming $D_{trm} = 0.0002$/day. (b) *assuming $t_{trm} = 9.7$ years.*

Solution
(a) First, the time to switch from hyperbolic to exponential decline is required:

$$t_{\text{trm}} = \frac{\frac{D_i}{D_{\text{trm}}} - 1}{bD_i} = \frac{\frac{0.022}{0.0002} - 1}{1.4 \times 0.022} = 3539 \text{ days} = 9.7 \text{ years}$$

Hyperbolic decline is to be used for the first 9.7 years. Therefore, the rate at the time of switching is:

$$q = q_i(1 + bD_i t)^{-1/b} = 7000(1 + 1.4 \times 0.022 \times 3539)^{-1/1.4} = 243.8 \text{ MSCF/day}$$

Therefore:

$$Q_{hyp} = \frac{q_i^b}{(1-b)D_i}\left(q_i^{1-b} - q^{1-b}\right) = \frac{7000^{1.4}}{(1-1.4)\times 0.022}\left(7000^{1-1.4} - 243.8^{1-1.4}\right)$$
$$= 2.25 \times 10^6 \text{ MSCF}$$

For the exponential decline:

$$Q_{exp} = \frac{q_i - q_{el}}{D} = \frac{243.8 - 10}{0.0002} = 1.17 \times 10^6 \text{ MSCF}$$

Therefore, the cumulative production is: $2.25 + 1.17 = 3.42$ BSCF.

Well life is the sum of the hyperbolic decline period and the period of exponential decline. The former is 9 years. The latter is obtained by:

$$t_{exp} = \frac{ln\,(q_i/q_{el})}{D} = \frac{ln\,(243.8/10)}{0.00014} = 15969 \text{ days} = 43.7 \text{ years}$$

Therefore, the well life is $(9.7 + 43.7) = 53.4$ years.

(b) The rate at the time of switching (9.7 years) is 243.8 STB/day as calculated above. Consequently, and again as shown above the cumulative production during the hyperbolic decline period is 2.25 BSCF. To determine the exponential decline cumulative production, the decline rate at which we switch from hyperbolic to exponential decline is required:

$$D_{trm} = \frac{D_i}{1 + bD_i t_{trm}} = \frac{0.022}{1 + 1.4 \times 0.022 \times 3539} = 0.0002/\text{day}$$

Therefore,

$$Q_{exp} = \frac{q_i - q_{el}}{D} = \frac{243.8 - 10}{0.0002} = 1.17 \times 10^6 \text{ MSCF}$$

Therefore, the cumulative production is: $2.25 + 1.17 = 3.42$ BSCF.

Well life is the sum of the hyperbolic decline period and the period of exponential decline. The former is 9.7 years. The latter is obtained by:

$$t_{exp} = \frac{ln\,(q_i/q_{el})}{D} = \frac{ln\,(243.8/10)}{0.0002} = 43.7 \text{ years}$$

Therefore, the total well life is 53.4 years.

Question 3.4 (True/False)
Arps' hyperbolic decline reduces to exponential decline when $b = 0$.

Question 3.5 (True/False)
The decline rate is always constant for Arps DCA.

Question 3.6 (True/False)
Major disadvantage of hyperbolic decline is that it results in unreasonable well life.

Question 3.7 (True/False)
Major disadvantage of hyperbolic decline is that it results in unreasonable well life when b-value is large.

Question 3.8 (True/False)

$$D(t) = \frac{-dq}{dQ}.$$

Question 3.9 (True/False)

$$b = \frac{d}{dt}\left(\frac{1}{D}\right).$$

3.4   Stretched Exponential DCA

Application of Arps DCA to shale/tight wells showed their inadequacy (Cheng et al. 2008; Ilk et al. 2008; Rushing et al. 2007; Spivey et al. 2001). DCA methods to address this shortcoming have been introduced since 2000s. Valko (2009) and Valkó and Lee (2010) introduced the stretched exponential decline method. The physical basis of the method was based on the assumption that the reservoir behavior can be considered as superposition of a great number of exponentially declining contributing volumes each showing different decline rate. The resulting equation is:

$$q = q_i \, exp\left(-\left(\frac{t}{\tau}\right)^n\right) \tag{3.15}$$

where τ is the characteristic time constant and n is the stretched exponential exponent. The stretched exponential has been shown to be applicable to exceptionally wide range of processes. Examples are materials relaxations in the materials science and engineering, urban population sizes in social sciences, and currency exchange rate fluctuations in economics (Elton 2018).

Integration of Eq. (3.15) results in the following equation for the cumulative production:

$$Q = \frac{q_i \tau}{n}\left[\Gamma\left(\frac{1}{n}\right) - \Gamma\left(\frac{1}{n}, \left(\frac{t}{\tau}\right)^n\right)\right] \tag{3.16}$$

where Γ is the Euler gamma function. The first term in the bracket represents the complete gamma function and the second term is incomplete gamma function. Assigning $q_{el} \approx 0$, the EUR is given by:

$$Q = \frac{q_i \tau}{n} \Gamma\left(\frac{1}{n}\right) \tag{3.17}$$

Well life is given by:

$$t = \tau \left[ln\left(\frac{q_i}{q}\right) \right]^{\frac{1}{n}} \tag{3.18}$$

The decline rate D and b exponent corresponding to hyperbolic decline can be calculated from the following:

$$D(t) = \frac{-1}{q(t)} \frac{dq}{dt} \approx n\tau^{-n} t^{n-1} \tag{3.19}$$

$$b = \frac{d}{dt}\left(\frac{1}{D}\right) = \frac{1-n}{n} \tau^n t^{-n} \tag{3.20}$$

Note that both D and b are time dependent considering the stretched exponential decline. This is unlike the hyperbolic decline where b was constant.

Example 3.3 $q_i = 10{,}000$ MSCF/day, $q_{el} = 10$ MSCF/day, $\tau = 0.3$ month, $n = 0.25$. Calculate well life and EUR.

Solution
First the well life is given by:

$$t = 0.6 \left[ln\left(\frac{10{,}000}{10}\right) \right]^{\frac{1}{0.25}} = 683.1 \text{ month} = 56.9 \text{ years}$$

The gamma function (programmed e.g. in Mathematica) should be used for calculation of EUR.

$$Q = \frac{10{,}000 \times 0.3 \times 30}{0.25} \left[\Gamma\left(\frac{1}{0.25}\right) - \Gamma\left(\frac{1}{0.25}, \left(\frac{683.1}{0.3}\right)^{0.25}\right) \right]$$

$$= 3.6e5[6 - 0.523] = 1.97e6 \text{ MSCF}$$

3.5 Power-Law Exponential DCA

Power law exponential was introduced based on the observed power-law behavior of D-parameter:

$$D = D_\infty + At^{-B} \tag{3.21}$$

Log-log plot of $D - D_\infty$ versus time shows linear behavior with negative slope. $D = D_\infty$ after long time when the boundary-dominated flow is established. Assuming small values of D_∞, the log-log plot of D versus t would show linear behavior. The above equation can be rewritten in the following form:

$$D = D_\infty + n\widehat{D}_i t^{-(1-n)} \tag{3.22}$$

Replacement into $D(t) = \frac{-1}{q(t)} \frac{dq}{dt}$ followed by integration gives:

$$q = \widehat{q}_i \, exp\left(-D_\infty t - \widehat{D}_i t^n\right) \tag{3.23}$$

Considering the definition $b = \frac{d}{dt}\left(\frac{1}{D}\right)$, it can be implied that b is time dependent too:

$$b = \frac{n\widehat{D}_i(1-n)}{\left[n\widehat{D}_i + D_\infty t^{(1-n)}\right]^2} t^{-n} \tag{3.24}$$

In some software implementations $\widehat{D}_i$ has been replaced by τ to relate the power-law DCA to stretched exponential DCA:

$$q = \widehat{q}_i \, exp\left(-D_\infty t - \left(\frac{t}{\tau}\right)^n\right) \tag{3.25}$$

This form of the equation clarifies that power-law DCA reduces to stretched exponential DCA when D_∞ is assumed zero. $\widehat{q}_i$ is rate intercept at $t = 0$ and should be considered a parameter of fit without assigning some physical meaning to it. It can be significantly larger the first recorded rate. n should be between 0 and 1. A range of 0.1–0.3 for n might be appropriate. Values of $\widehat{D}_i$ might be from 0.5 to 3/day. D_∞ can be as low as zero but will most likely be in the 10^{-4} to 10^{-3}/day range.

Note that there is no closed-form cumulative production equation for this technique. $D(t) = \frac{-1}{q(t)} \frac{dq}{dt} = \frac{-dq}{dQ}$ should be log-log plotted versus time. Once straight-line is identified on log-log plot of D versus t, slope and intercept are obtained from which D_i and n are calculated. q_i is obtained by matching semilogy plot of rate versus time.

D_∞ converts the power-law exponential equation to an exponential decline eventually. However, D_∞ is not required unless D is observed to be converging to a constant value.

Example 3.4 $q_i = 800{,}000$ MSCF/day, $\widehat{D}_i = 4$/day, $D_\infty = 0.000464$/day, $n = 0.07$, $q_{el} = 10$ MSCF/day. Calculate well life and EUR.

Solution

$$q = \widehat{q}_i\, exp\left(-D_\infty t - \widehat{D}_i t^n\right) \rightarrow ln\left(\frac{800{,}000}{10}\right) = 0.000464t + 4t^{0.07}$$

Therefore, $t = 8140$ days or 22.3 years. Using the numerical integration, the EUR is 2.82×10^6 MSCF.

3.6 Duong DCA

Duong model requires long-term linear flow for which the log-log plot of rate/cumulative production vs. time yields a negative unit slope straight line:

$$\frac{q}{Q} = at^{-m} \tag{3.26}$$

Introducing the following rate-time relationship:

$$q(t) = q_i t^{-m}\, exp\left[\frac{a}{(1-m)}\left(t^{(1-m)} - 1\right)\right] \tag{3.27}$$

This leads to the following cumulative rate equation:

$$Q = \frac{q_i}{a}\, exp\left[\frac{a}{(1-m)}\left(t^{(1-m)} - 1\right)\right] \tag{3.28}$$

This model could yield optimistic EUR and well life values. The well life can be obtained based on the rate-time relationship assuming a given economic limit rate. The EUR will be next calculated based on:

$$Q = \frac{q_{el}}{at^{-m}} \tag{3.29}$$

Example 3.5 $a = 1.42$/day, $m = 1.14$, $q_i = 19{,}000$ m³/day, $q_{el} = 1000$ m³/day. Calculate EUR and well life.

Solution
Based on the time-rate equation, we can write:

$$4 = 19{,}000t^{-1.14} \, exp \left[\frac{1.42}{(1 - 1.14)} \left(t^{(1 - 1.14)} - 1 \right) \right] \rightarrow t = 7573 \, \text{day} = 20.7 \, \text{years}$$

EUR is given by:

$$\text{EUR} = \frac{1000}{1.42 \times 7572^{-1.14}} = 1.86 \times 10^7 \text{m}^3$$

3.7 Logistic Growth DCA

The cumulative production for this model is given by:

$$Q = \frac{Kt^n}{a + t^n} \tag{3.30}$$

K is the carrying capacity. Note that $Q \rightarrow K$ when $t \rightarrow \infty$. In other words, K can be considered the EUR with no economic limit constraint. n controls the steepness of the decline. Higher n values exhibit more gradual decline. Rate declines faster for lower a values. a corresponds to the time at which half of the carrying capacity is reached because setting $t^n = a$ gives $Q = K/2$. The rate-time relationship is given by differentiating the above equation which gives:

$$q = \frac{Knat^{n-1}}{(a + t^n)^2} \tag{3.31}$$

Also,

$$D = \frac{\frac{2an}{a+t^n} - (1 + n)}{t} \tag{3.32}$$

Example 3.6 $K = 328{,}000$ STB, $n = 0.69$, $a = 151.7$ month^n. Calculate terminal rate and EUR assuming 30-year well life.

Solution

$$q = \frac{Knat^{n-1}}{(a+t^n)^2} = \frac{328{,}000 \times 0.69 \times 151.7 \times 30^{0.69-1}}{\left(151.7 + 30^{0.69}\right)^2} = 156.6 \text{ STB/day}$$

And

$$Q = \frac{Kt^n}{a+t^n} = \frac{328{,}000 \times 30^{0.69}}{151.7 + 30^{0.69}} = 173{,}500 \text{ STB}$$

3.8   Graphical Visualization

In order to identify the suitable DCA method and find the corresponding parameters for a given Q-q-t dataset, different graphical visualizations can be used. The graphs may be Cartesian, semi-log, or log-log. In addition to q, Q, t, one can plot D, b, q/Q and other functions. Specialized graphs for each DCA method has been also proposed. However, the following four plots are widely cited as generally useful (e.g. see Houze et al. 2020).

1. Semilogy of Q and q versus t: This should be the primary graph to decide the goodness of fit. Linear behavior of log (q) versus t indicates Arps' exponential decline. Non-linear behavior is expected for shale wells.
2. log-log plot of D versus t: Constant D indicates Arps' exponential decline. Straight-line behavior can indicate Arps' hyperbolic decline if b is constant.
3. log-log plot of b versus t: Constant b indicates Arps' hyperbolic decline. Variable b requires non-Arps or segmented (switching from hyperbolic to exponential) Arps DCA.
4. log-log of q/Q versus t: Linear behavior shows linear flow regime is dominant. Consequently, Duong DCA should be suitable. A reminder that Duong DCA is generally optimistic.

Question 3.10
The physical basis of the _____ DCA method was based on the assumption that the reservoir behavior can be considered as superposition of a great number of exponentially declining contributing volumes each showing different decline rate

(a) Arps' exponential
(b) Power law exponential
(c) Stretched exponential
(d) Logistic growth

Question 3.11
Which of the following DCA techniques does not have a closed-form cumulative production equation?

(a) Arps' exponential
(b) Power law exponential
(c) Stretched exponential
(d) Logistic growth

Question 3.12
b-value is constant for which of the following DCA techniques?

(a) Power law exponential
(b) Stretched exponential
(c) Logistic growth
(d) Neither

Questions Answers
3.1. a
3.2. d
3.3. $\frac{q}{q_i} = e^{-Dt} = e^{-1} = 0.37$
3.4. True
3.5. False
3.6. False
3.7. True
3.8. True
3.9. True.
3.10. c
3.11. b
3.12. d

Chapter 4
Rate-Pressure-Time Analysis for Reserve Estimation

Abstract In departure from the simple rate-time decline curve analysis (DCA) methods presented in the previous chapter, in this chapter, we present the advanced techniques bringing the pressure into the analysis. DCA methods are essentially empirical introduced based on observation of data behavior. While the rate-pressure-time analysis is physics-based and introduced based on solving the fundamental equations governing flow in the reservoir. The rate-pressure-time analysis is often referred to as rate transient analysis (RTA). In this chapter, the fundamentals of RTA as applied to conventional wells is first introduced. Next, log-log analysis procedure are introduced accordingly along with the flowing material balance plot. The analysis is next extended to conventional gas reservoirs. Adapting the RTA methods to shale wells requires understanding the flow regimes in multi-fractured horizontal wells and their corresponding analysis techniques. As a result, the conventional RTA methods are extended to shale wells by introducing analysis approaches of transient linear flow, transitional flow, and stimulated-reservoir-volume (SRV) flow. The introduced methods are applied to example problems throughout the chapter.

4.1 RTA for Conventional Vertical Wells

A good starting point to introduce RTA is the analytical solution to the diffusivity equation subject to constant pressure production from a vertical well in a bounded circular reservoir:

$$\bar{q}_D = \frac{1}{\sqrt{S}} \frac{K_1\left(\sqrt{S}\right) I_1\left(r_D\sqrt{S}\right) - I_1\left(\sqrt{S}\right) K_1\left(r_D\sqrt{S}\right)}{K_0\left(\sqrt{S}\right) I_1\left(r_{eD}\sqrt{S}\right) - I_0\left(\sqrt{S}\right) K_1\left(r_{eD}\sqrt{S}\right)} \tag{4.1}$$

Where

© The Author(s), under exclusive license to Springer Nature Switzerland AG 2023
M. Zeidouni, *Shale Hydrocarbon Recovery*, SpringerBriefs in Earth Sciences,
https://doi.org/10.1007/978-3-031-23559-7_4

$$q_D = \frac{q\mu B}{2\pi kh \left(p_i - p_{wf}\right)} \tag{4.2}$$

$$t_D = \frac{\eta t}{r_{wa}^2} \tag{4.3}$$

$$r_{eD} = \frac{r_e}{r_{wa}} \tag{4.4}$$

$$r_{wa} = r_w e^{-s} \tag{4.5}$$

In the above equation, q is rate, r is radius from the wellbore, and t is time. S is the Laplace domain dummy variable. μ is viscosity, B is the fluid formation volume factor, k is permeability, h is reservoir thickness, and s is skin factor. r_w, r_{wa}, and r_e represent the wellbore radius, apparent wellbore radius, and reservoir external radius, respectively. p_i and p_{wf} indicate the initial pressure and well flowing pressure, respectively. η is the diffusivity coefficient $\left(= \frac{k}{\phi\mu c_i}\right)$. The subscript D indicates the dimensionless properties. K_0 and K_1 are the modified Bessel functions of the second kind of order zero and one, respectively.

At early-time the flow regime is radial transient flow. After some time, the flow becomes boundary-dominated flow (BDF). Shortly after BDF is established, the above solution reduces to the following asymptotic constant-pressure solution:

$$q_D = \frac{1}{\ln\left(r_{eD}\right) - \frac{1}{2}} \, exp \left(\frac{-t_D}{\frac{1}{2}\left(r_{eD}^2 - 1\right)\left(\ln\left(r_{eD}\right) - \frac{1}{2}\right)}\right) \tag{4.6}$$

This equation rigorously provides the fundamental basis for the empirically derived Arps' exponential decline ($q = q_i \exp(-Dt)$). Comparison of the asymptotic solution with the Arps' exponential decline gives:

$$q_i = \frac{2\pi kh \left(p_i - p_{wf}\right)}{\mu B \left(\ln\left(\frac{r_e}{r_{wa}}\right) - \frac{1}{2}\right)} \tag{4.7}$$

$$D = \frac{\frac{\eta}{r_{wa}^2}}{\frac{1}{2}\left(\left(\frac{r_e}{r_{wa}}\right)^2 - 1\right)\left(\ln\left(\frac{r_e}{r_{wa}}\right) - \frac{1}{2}\right)} \tag{4.8}$$

Note that this solution is strictly for constant-pressure boundary condition at the well. Theoretically speaking, exponential decline should be only applicable when pressure is constant. However, (1) this assumption was not a requirement when Arps DCA was introduced, and (2) this assumption is rarely valid in practice. In practice, both pressure and rate are generally variable. An equation that is valid regardless of pressure and rate variations is required. The two extreme cases to be considered are that either the pressure is fixed and rate varies or the rate is fixed and the pressure

Fig. 4.1 Top view of the closed circular reservoir

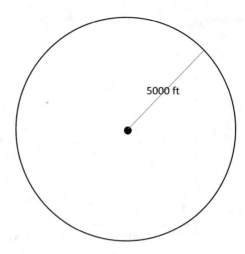

5000 ft

Table 4.1 Reservoir rock, fluid, and well information for the base case oil well

Well radius, r_w	0.3	ft
Pay zone, h	30	ft
Total compressibility, c_t	3×10^{-6}	1/psi
Porosity, ϕ	0.1	–
Top reservoir depth	6000	ft
Initial pressure, p_i	5000	psi
Temperature, T	212	F
Formation volume factor, B_o	1	bbl/STB
Viscosity, μ	1	cp
Water saturation, S_w	0	–
Production period, t	108,720	h
Bottom-hole pressure, p_{wf}	750	psi
Permeability, k	33.33	mD
Reservoir radius, r_e	6000	ft
Skin factor, s	5	–

varies. It is essential to obtain an equation that equally applies to both of these end cases.

Consider a closed circular reservoir (Fig. 4.1) with rock and fluid properties and well information provided in Table 4.1. The reader is recommended to numerically simulate the production for this base case. The author used Kappa-Rubis (2020) for this purpose. The flow rate versus time is shown in Fig. 4.2. Note that the late time clearly shows linear behavior indicating exponential decline, as expected. Figure 4.3 shows the same rate versus time data plotted in log-log graph.

Alternatively, the rate can be plotted versus the material balance time, $t_{MB} = Q/q$. The material balance time would be identical to the actual time if flow rate is constant. When flow rate declines, t_{MB} must be increasing with time. Figure 4.4 shows t_{MB} and t_{MB}/t versus time on log-log graph. Blasingame et al. (1991) showed

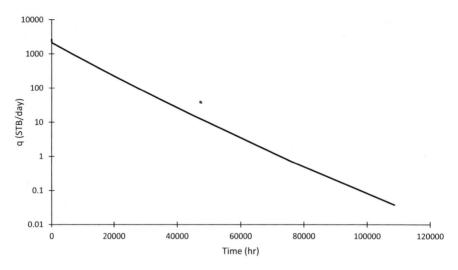

Fig. 4.2 Log (q) versus time illustrating the exponential rate decline for constant pressure BDF

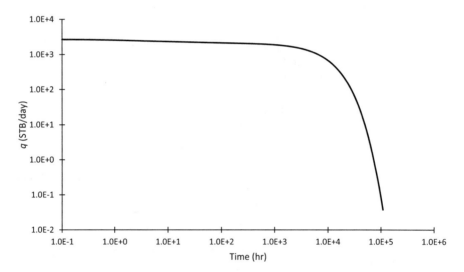

Fig. 4.3 Log-log plot of q versus time

that if the constant-pressure rate q (or $q/\Delta p$ since Δp is constant for constant pressure) is plotted versus the material balance time, the resulting log-log graph would show negative unit-slope at the late time (Fig. 4.5). More importantly the solution will overlap the constant-rate solution as shown in Fig. 4.6.

The solution overlap can be shown in terms of $\Delta p/q$ (or rate-normalized pressure, RNP) versus t_{MB} which shows unit-slope at late-time (Fig. 4.7). Figure 4.7 also shows the overlapping constant-rate and constant-pressure responses in terms of log-derivative of RNP. Figure 4.7 should remind us of the log-log diagnostic plot in

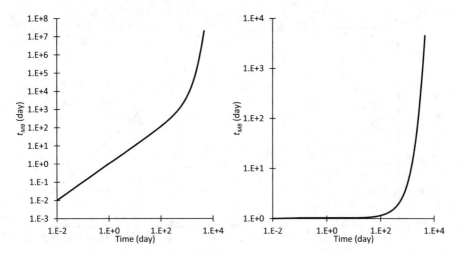

Fig. 4.4 Material balance time versus time

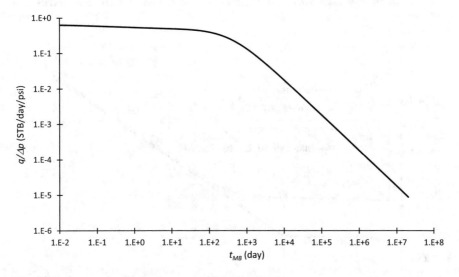

Fig. 4.5 Pressure normalized rate versus material balance time

drawdown pressure transient analysis where the flow rate is constant. Given the overlapping behavior, the drawdown pressure transient analysis (PTA) methodology can be readily extended to constant-pressure rate analysis. Note the zero-slope during transient period and unit-slope during the late-time BDF. We refer to this plot as log-log (diagnostic) plot.

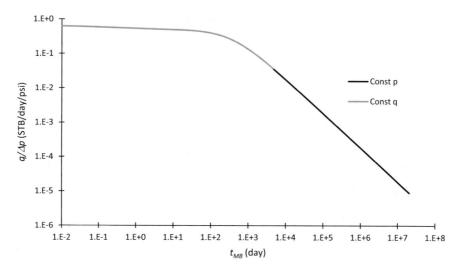

Fig. 4.6 Overlapping $q/\Delta p$ versus t_{MB} responses for constant-rate and constant-pressure well conditions

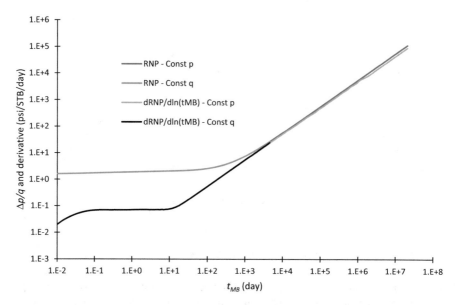

Fig. 4.7 Overlapping constant-pressure and constant-rate responses in terms of RNP and its derivative versus t_{MB}

4.1.1   Log-Log Plot Analysis

We can now use the same diagnostic plot used in drawdown well testing for RTA. The resulting diagnostic plot includes $\Delta p/q$ and its derivative. This log-log plot is one

of the key plots used in RTA. From late-time data we can obtain the reservoir size, and from early time we can obtain permeability or flow capacity (kh) and skin factor. To use the transient radial flow period, the following equation is known from PTA:

$$p_i - p_{wf} = \frac{162.6q\mu B}{kh}\left[\log(t) + \log\left(\frac{k}{\phi\mu c_t r_w^2}\right) - 3.23 + 0.869s\right] \tag{4.9}$$

The time unit in this equation is hours. Given the long times of interest in RTA compared to PTA, the time unit should be changed from hours to days. The resulting equation becomes:

$$p_i - p_{wf} = \frac{162.6q\mu B}{kh}\left[\log(t) + \log\left(\frac{k}{\phi\mu c_t r_w^2}\right) - 1.85 + 0.869s\right] \tag{4.10}$$

Remember that from the value of zero-slope derivative corresponding to transient radial flow (m'), we get:

$$m' = \frac{d(\Delta p/q)}{d\ln(t)} = \frac{d(\Delta p/q)}{2.303\,d\log(t)} = \frac{1}{2.303}\frac{162.6\mu B}{kh} = \frac{|m|}{2.303} \tag{4.11}$$

where m is the slope of the semi-log graph of $\Delta p/q$ versus t. Permeability can then be obtained by:

$$k = \frac{1}{2.303}\frac{162.6\mu B}{hm'} \tag{4.12}$$

From the intercept of semi-log graph of $\Delta p/q$ versus t, i.e. $(\Delta p/q)_{1\ day}$ we get:

$$s = 1.151\left[\frac{(\Delta p/q)_{1day}}{\frac{162.6\mu B}{kh}} - \log\left(\frac{k}{\phi\mu c_t r_w^2}\right) + 1.85\right] \tag{4.13}$$

Or

$$s = 1.151\left[\frac{(\Delta p/q)_{1day}}{2.303m'} - \log\left(\frac{k}{\phi\mu c_t r_w^2}\right) + 1.85\right] \tag{4.14}$$

The boundary-dominated flow under constant production rate is often referred to as pseudo-steady state flow (PSSF). The PSSF solution can be extended to constant pressure. The constant rate solution is given by (see derivation in Appendix 1):

$$\frac{p_i - p_{wf}}{q} = \frac{\mu B}{2\pi kh}\left[\frac{1}{2}\ln\left(\frac{4A}{e^{\gamma}C_A r_w^2}\right) + s\right] + \frac{B(1 - S_{wi})}{NB_i c_t}t \qquad (4.15)$$

Replacement of time by material balance time makes this equation applicable to constant pressure:

$$\frac{p_i - p_{wf}}{q} = \frac{\mu B}{2\pi kh}\left[\frac{1}{2}\ln\left(\frac{4A}{e^{\gamma}C_A r_w^2}\right) + s\right] + \frac{B(1 - S_{wi})}{NB_i c_t}t_{MB} \qquad (4.16)$$

In field units (material balance time is in days):

$$\frac{p_i - p_{wf}}{q} = \frac{141.2\mu B}{kh}\left[\frac{1}{2}\ln\left(\frac{4A}{e^{\gamma}C_A r_w^2}\right) + s\right] + \frac{B(1 - S_{wi})}{NB_i c_t}t_{MB} \qquad (4.17)$$

In short, this can be written as:

$$\frac{\Delta p}{q} = \frac{p_i - p_{wf}}{q} = b_{pss} + m_{pss}t_{MB} \qquad (4.18)$$

where

$$m_{pss} = \frac{1 - S_{wi}}{Nc_t}\frac{B}{B_i} \qquad (4.19)$$

$$b_{pss} = \frac{141.2\mu B}{kh}\left[\frac{1}{2}\ln\left(\frac{4A}{e^{\gamma}C_A r_w^2}\right) + s\right] \qquad (4.20)$$

After sufficiently long time, b_{pss} becomes negligible compared to $m_{pss}t_{MB}$. Therefore, the late-time data on the log-log plot will follow a unit-slope line with intercept of $\frac{B(1 - S_{wi})}{NB_i c_t}$ from which N can be obtained.

Example 4.1 Calculate permeability, skin factor, and oil initial in place for the data shown in Fig. 4.7.

Solution
The value of zero-slope line is 0.07 STB/day/psi. To obtain $(\Delta p/q)_{1\ day}$, the semi-log graph of RNP data corresponding to transient radial flow should be plotted versus time (Fig. 4.8). The intercept of the fitted line with the y-axis at 1 day is $(\Delta p/q)_{1\ day}$ which is 1.877 STB/day/psi. In addition, the intercept of the late-time line-fitted derivative at 1 day is 5.5×10^{-3} psi/STB/day.

From value of zero-slope line (0.07 STB/day/psi), we can obtain the permeability:

$$k = \frac{162.6\mu B}{2.303m'h} = \frac{162.6 \times 1 \times 1}{2.303 \times 0.07 \times 30} = 33.6\ \text{mD}$$

The actual permeability is 33.33 mD.

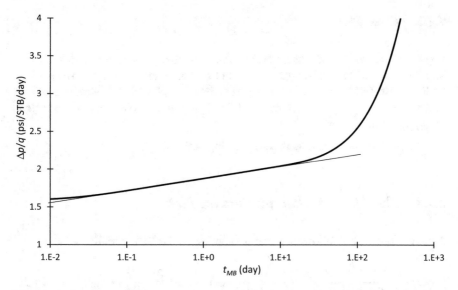

Fig. 4.8 Semi-log graph of RNP versus time. Line-fitting the data corresponding to radial transient flow gives $(\Delta p/q)_{1\ \text{day}}$

$$s = 1.151\left[\frac{(\Delta p/q)_{1hr}}{2.303m'} - \log\left(\frac{k}{\phi\mu c_t r_w^2}\right) + 1.85\right]$$

$$= 1.151\left[\frac{1.877}{2.303 \times 0.07} - \log\left(\frac{32.7}{0.1 \times 1 \times 3 \times 10^{-6} \times 0.3^2}\right) + 1.85\right] = 5.07$$

The actual skin factor is 5.

$$\frac{B(1 - S_{wi})}{NB_i c_t} = \frac{1}{Nc_t} = 5.5 \times 10^{-3} \rightarrow N = \frac{1}{3 \times 10^{-6} \times 5.5 \times 10^{-3}} = 6.063 \times 10^{7}\ \text{STB}$$

The actual N is given by:

$$N = \frac{Ah\phi(1 - S_{wi})}{B_{oi}} = \frac{\pi(6000^2)(30)(0.1,\ 1)}{1} = 3.39 \times 10^{8}\ \text{ft}^3 = 6.04 \times 10^{7}\ \text{STB}$$

In addition to the log-log plot above, two other figures are plotted as discussed in the following.

4.1.2   PNR Plot

Log-log PNR plot is sometimes preferred since it readily visualizes the rate decline. The PNR derivative will be negative and not shown on log-log graph. Instead of the PNR and its derivative, the following functions are plotted: (1) Normalized PNR integral: $I_{PNR} = \frac{1}{t_{MB}} \int_0^{t_{MB}} PNR(\tau)d\tau$, and (2) Normalized PNR integral derivative: $I_{PNR}\text{derivative} = -\frac{d(PNR\text{int.})}{d \ln t_{MB}}$

This plot for the base case is shown in Fig. 4.9.

4.1.3   The Flowing Material Balance Plot

As noted earlier, the following equation applies for the BDF period:

$$\frac{p_i - p_{wf}}{q} = \frac{141.2\mu B}{kh}\left[\frac{1}{2}\ln\left(\frac{4A}{e^\gamma C_A r_w^2}\right) + s\right] + \frac{B(1 - S_{wi})}{NB_i c_t}t_{MB} \tag{4.21}$$

Replacement of t_{MB} by Q/q gives:

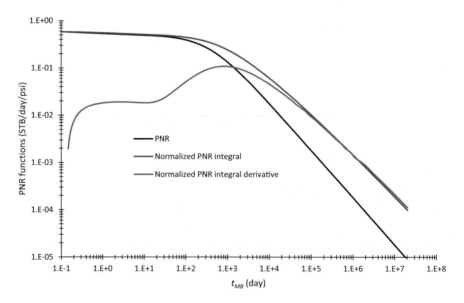

Fig. 4.9 PNR and related functions for the base case

$$\frac{p_i - p_{wf}}{q} = \frac{141.2\mu B}{kh}\left[\frac{1}{2}\ln\left(\frac{4A}{e^\gamma C_A r_w^2}\right) + s\right] + \frac{B(1-S_{wi})}{NB_i c_t}\frac{Q}{q} \tag{4.22}$$

Multiplying both sides by $q/(p_i\text{-}p_{wf})$ gives:

$$1 = \frac{q}{\Delta p}\frac{141.2\mu B}{kh}\left[\frac{1}{2}\ln\left(\frac{4A}{e^\gamma C_A r_w^2}\right) + s\right] + \frac{B(1-S_{wi})}{NB_i c_t}\frac{Q}{\Delta p} \tag{4.23}$$

Rearrangement gives:

$$\frac{q}{\Delta p} = \frac{1}{\frac{141.2\mu B}{kh}\left[\frac{1}{2}\ln\left(\frac{4A}{e^\gamma C_A r_w^2}\right) + s\right]}$$
$$- \frac{B(1-S_{wi})}{NB_i}\frac{1}{\frac{141.2\mu B}{kh}\left[\frac{1}{2}\ln\left(\frac{4A}{e^\gamma C_A r_w^2}\right) + s\right]}\frac{Q}{c_t\Delta p} \tag{4.24}$$

Based on the above equation, plotting $q/\Delta p$ versus $Q/(c_t\,\Delta p)$ should exhibit linear behavior during BDF. The slope and intercept of the plot can be used to obtain N:

$$\frac{B(1-S_{wi})}{NB_i} = \frac{\text{slope}}{\text{intercept}} \tag{4.25}$$

For example, for the flowing material balance plot shown in Fig. 4.10 corresponding to the base case, we get slope/intercept $= 1.66 \times 10^{-8}$ and therefore, $N = 6.03 \times 10^7$ STB. The actual N is 6.07×10^7 STB as noted above.

4.1.4   Extension to Gas Reservoirs

The same diagnostic plots as above can be applied to gas following some modifications. The pressure should be replaced by the adjusted pseudo-pressure defined by:

$$p_a = \frac{\mu_i z_i}{p_i}\int_{P_{ref}}^{P}\frac{p}{\mu z}\,dp \tag{4.26}$$

Also, the cumulative production should be replaced by adjusted pseudo-Q:

$$Q_a = \int_0^t \frac{q\mu_i c_{ti}}{\mu c_t}\,dt \tag{4.27}$$

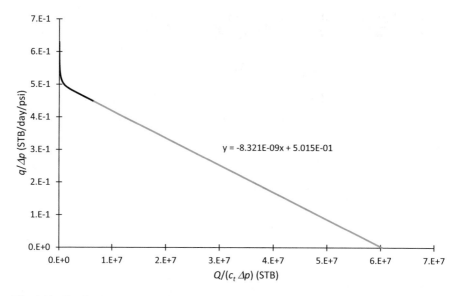

Fig. 4.10 The flowing material balance plot for the base case

The bar sign indicates viscosity and total compressibility evaluated at the average reservoir pressure. The adjusted pseudo material balance time should then be calculated by:

$$t_{a,MB} = \frac{Q_a}{q} \tag{4.28}$$

The permeability can then be obtained from the value of zero-slope derivative corresponding to transient radial flow (m'):

$$k = \frac{1}{2.303} \frac{162.6\mu_i B_{gi}}{hm'} \tag{4.29}$$

where

$$m' = \frac{d(\Delta p_a/q)}{d \ln t_{a,MB}} \tag{4.30}$$

And the skin factor is obtained by:

$$s = 1.151 \left[\frac{(\Delta p_a/q)_{1day}}{2.303m'} - \log\left(\frac{k}{\phi\mu_i c_{ti} r_w^2}\right) + 1.85 \right] \tag{4.31}$$

Similar to the oil case, the late-time data on the log-log plot will follow a unit-slope line with intercept of $\frac{(1-S_{wi})}{Gc_{ti}}$ from which G can be obtained. This is based on the modified BDF equation given by:

$$\frac{\Delta p_a}{q} = \frac{p_{a,i} - p_{a,wf}}{q} = b_{pss} + m_{pss}t_{a,MB} \tag{4.32}$$

Where

$$m_{pss} = \frac{1 - S_{wi}}{Gc_{ti}} \tag{4.33}$$

$$b_{pss} = \frac{141.2\mu_i B_i}{kh}\left[\frac{1}{2}\ln\left(\frac{4A}{e^\gamma C_A r_w^2}\right) + s\right] \tag{4.34}$$

After sufficiently long time, b_{pss} becomes negligible compared to $m_{pss}t_{a,\ MB}$. Therefore, the late-time data on the log-log plot will follow a unit-slope line with intercept of $\frac{(1-S_{wi})}{Gc_{ti}}$ from which G can be obtained.

The flowing material balance based on the above equation would be the plot of q/dp_a versus $Q_d/(c_{ti} \times \Delta p_a)$. The slope and intercept of line-fitted BDF data would give:

$$\frac{(1-S_{wi})}{G} = \frac{\text{slope}}{\text{intercept}} \tag{4.35}$$

Example 4.2 Consider a base case dry gas well with properties listed in the Table 4.2. Gas rate and cumulative production are shown versus time in Fig. 4.11.

Solution
The log-log plot corresponding to the base case gas well is shown in Fig. 4.12.

Table 4.2 Reservoir rock, fluid, and well information for the base case gas well

Well radius, r_w	0.3	ft
Pay zone, h	30	ft
Rock compressibility, c_t	3×10^{-6}	1/psi
Porosity, ϕ	0.1	–
Top reservoir depth	6000	ft
Initial pressure, p_i	5000	psi
Temperature, T	212	F
Water saturation, S_w	0.3	–
Production period, t	1	year
Bottom-hole pressure, p_{wf}	1000	psi
Permeability, k	10	mD
Reservoir radius, r_e	3000	ft
Skin factor, s	5	–
Production period	10	years

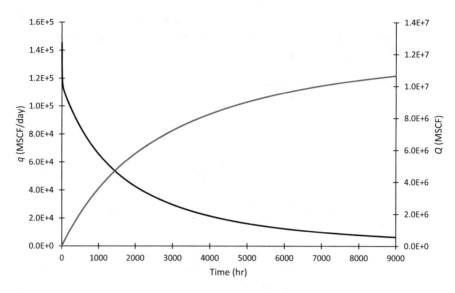

Fig. 4.11 Rate and cumulative gas production versus time for the base case gas well

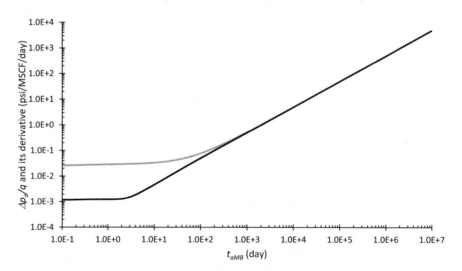

Fig. 4.12 The adjusted pseudo-RNP and its derivative versus adjusted pseudo-t_{MB} for the base case gas well

According to this figure m' = 0.0012 psi/MSCF/day = 6.738 × 10^{-6} psi/STB/day. Therefore,

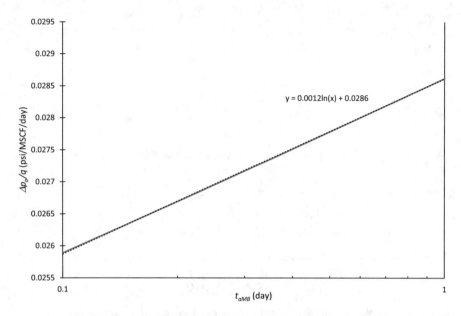

Fig. 4.13 The adjusted pseudo-RNP versus adjusted pseudo-t_{MB} for the base case gas well data during the transient radial flow period

$$k = \frac{1}{2.303}\frac{162.6\mu_i B_{gi}}{hm'} = \frac{1}{2.303}\frac{162.6 \times 0.02653 \times 0.003817}{30 \times 6.738 \times 10^{-6}} = 35.4 \text{ mD}$$

The actual permeability was 33.33 mD. The semi-log graph of $\Delta p_a/q$ versus t_{aMB} for the transient period's data is required to obtain $(\Delta p_a/q)_{1 \text{ day}}$ (Fig. 4.13). Based on this figure, $(\Delta p_a/q)_{1 \text{ day}} = 0.0286$ psi/MSCF/day $= 0.00016$ psi/STB/day. Therefore,

$$s = 1.151\left[\frac{0.00016}{2.303 \times 6.738 \times 10^{-6}} - \log\left(\frac{35.4}{0.1 \times 0.02653 \times 9.05 \times 10^{-5} \times 0.3^2}\right)\right.$$
$$\left. + 1.85\right]$$
$$= 3.44$$

The actual skin factor is 5.

The intercept of unit-slope line is 0.0005 psi/Mscf. Therefore,

$$\frac{(1 - S_{wi})}{Gc_{ti}} = \frac{(1 - 0.3)}{G(9.05 \times 10^{-3})} = 5 \times 10^{-4} \rightarrow G = 1.55 \times 10^7 \text{ MSCF}$$

This is identical to the actual G. The FMB plot is shown in Fig. 4.14 according to which G is obtained by:

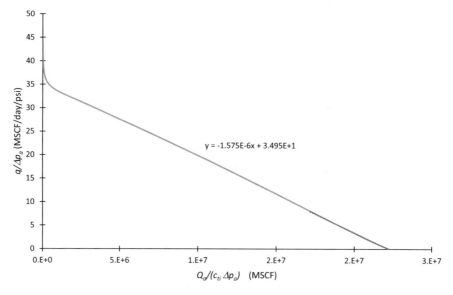

Fig. 4.14 The FMB plot for the base case gas well data line-fitted for the BDF period

$$\frac{(1 - S_{wi})}{G} = \frac{\text{slope}}{\text{intercept}} \rightarrow \frac{(1 - 0.3)}{G} = \frac{1.575 \times 10^{-6}}{34.95} \rightarrow G = 1.55 \times 10^{7} \text{ MSCF}$$

4.2   Extension to Unconventional Shale Wells

When introducing RTA for vertical wells, the major flow regimes of interest are the transient radial flow and BDF as discussed above. For horizontal wells and multi-fractured horizontal wells (MFHWs) used for shale hydrocarbon recovery, different flow regimes are observed. In the following, we first introduce the flow regimes observed in MFHWs. Next, we present the RTA approach specific to MFHWs.

4.2.1   MFHWs' Flow Regimes

At the beginning of production, the flow is orthogonal to the fractures (linear). Because of very low permeability of the reservoir matrix, the pressure depletion caused by one fracture is not felt by neighboring fracture during this flow regime. The reservoir acts as if it sees a single very large fracture with surface area equal to the sum of the surface areas of all fractures. This period of transient linear flow can be used to obtain information on the reservoir permeability combined with fractures' length.

The transient linear flow terminates after the pressure drops caused by neighboring fractures start to interfere. This period of pressure interference between fractures is referred to as transitional flow.

The average permeability of the fractured reservoir volume or the stimulated reservoir volume (SRV) is much larger than the original reservoir's permeability prior to fracturing. One may expect that the system would act as a composite infinite-acting reservoir after the SRV boundary is reached. Consequently, a new transient linear flow would be expected. However, the significant contrast between the permeability of the inner SRV and outer original reservoir causes the outer reservoir to be seen as (pseudo-)closed boundary. This is similar to the behavior observed in gas reservoirs supported by edge water influx. The pseudo-BDF observed when the pressure effect reaches the boundary of the SRV is referred to as SRV flow. Being a BDF flow, the SRV flow is characterized by a close to unity slope on the log-log graph the analysis of which can provide information on the size of the SRV. To model the SRV flow when analyzing data the permeability of the reservoir around the fractures may need to be enhanced (see the base case MFHW model in next section).

As time progresses, more of the unstimulated reservoir gets depleted which would eventually leads to infinite-acting radial flow if there is no neighboring well or boundary. In practice, radial flow is never reached. If this flow would have been established, the permeability could have been estimated from zero-slope value of the derivative on the log-log plot. Figure 4.15 Shows the abovementioned sequence of flow regimes.

4.2.2   MFHWs' Flow Regimes Identification and Analysis

In short, during production of MFHWs at least three flow periods can be identified in chronological order: (1) linear, (2) transitional, and (3) SRV flow. These three flow periods are shown in the log-log plot (Fig. 4.16) for the properties given in Table 4.3 for a base case MFHW model.

Linear Flow

On the log-log plot, linear flow is identified by a half-slope on both $\Delta p/q$ and its derivative where the two half-slope lines are separated by a factor 2 (Fig. 4.16). The pressure distribution during this flow is shown in Fig. 4.17.

Information on surface area from this linear flow regime is obtained from square-root time plot. The equation relating the flow rate to pressure change for linear flow under constant rate is given by (see derivation is Appendix 2):

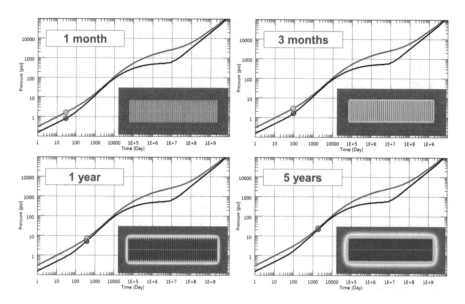

Fig. 4.15 Flow regimes in a MFHW: (**a**) transient linear flow, (**b**) transitional flow, (**c**) SRV flow, (**d**) transient radial flow. Note the reservoir top view shown as sub-plot in each panel shows the pressure distribution. The reservoir size is chosen very large so that the pressure never reaches its boundaries. The reservoir top view is zoomed on the SRV region. © KAPPA Eng (Houze et al. 2022)

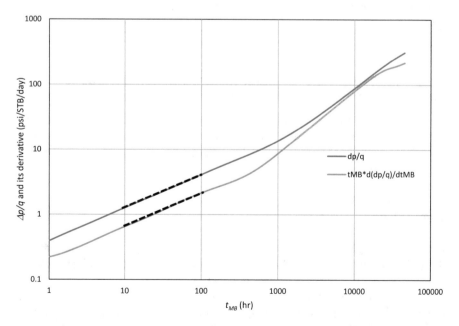

Fig. 4.16 Log-log plot showing the three flow periods in a MFHW. The black dashed lines indicate linear flow (0.5-slope). The green dashed line shows the SRV flow (unit-slope)

Table 4.3 Properties used in illustrating MFHW's flow regimes

Matrix porosity	–	0.05
Reservoir length	ft	10000
Reservoir width	ft	10000
Initial pressure	psia	8700
Reservoir temperature	F	200
Oil API gravity	–	42
Initial Water Saturation*	–	0
Fracture Height	ft	200
Net Pay	ft	200
Depth	ft	11000
Lateral length	ft	1000
Fracture Half length	ft	140
Number of fractures	–	20
Reservoir permeability	mD	4×10^{-5}
Well flowing pressure	psi	3000
rock compressibility	/psi	3×10^{-6}
Oil compressibility	/psi	1.66912×10^{-5}
Formation volume factor at 8000 psi	bbl/STB	2.2
Oil viscosity	cp	0.2
Enhanced permeability around fractures	mD	4×10^{-4}

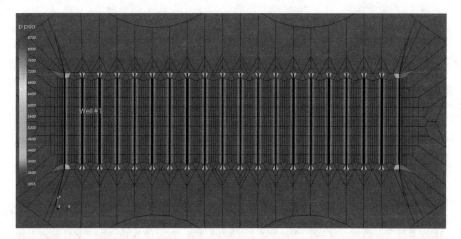

Fig. 4.17 The pressure distribution in the reservoir during early-time linear flow at 40 h ($t_{MB} = 80.6$ h)

$$\frac{p_i - p_{wf}}{q} = 19.89 \frac{B}{h x_f} \sqrt{\frac{\mu}{\phi c_t k}} \sqrt{t} \qquad (4.36)$$

This equation subject to constant flowing pressure is (see derivation is Appendix 3):

$$\frac{p_i - p_{wf}}{q} = 31.3 \frac{B}{hx_f} \sqrt{\frac{\mu}{\phi c_t k}} \sqrt{t} \qquad (4.37)$$

It may be more relevant to use the constant-pressure solution in RTA. However, the constant rate solution has been used in some software applications (e.g. Kappa and IHS). The constant pressure solution is used herein. Based on the constant pressure equation, plotting $\Delta p/q$ versus $\sqrt{t}$ can be line-fitted with the slope:

$$\alpha = \frac{31.3B}{hx_f} \sqrt{\frac{\mu}{\phi c_t k}} \qquad (4.38)$$

Therefore,

$$x_{mf}\sqrt{k} = \frac{31.3B}{\alpha h} \sqrt{\frac{\mu}{\phi c_t}} \qquad (4.39)$$

Or

$$n_f x_f \sqrt{k} = \frac{31.3B}{\alpha h} \sqrt{\frac{\mu}{\phi c_t}} \qquad (4.40)$$

Transitional Flow (from Linear to SRV)

After some time, pressure disturbances caused by neighboring fractures start to interfere causing the slope to become larger than that for linear flow (1/2). This is the transitional flow regime the pressure behavior during which is shown in Fig. 4.18. Transitional flow may be observed for a significant time in MFHW production. This is mainly due to the non-uniformity in the fractures' lengths and spacing originating from the complexity of the created fracture network.

SRV Flow

After full interference between the fractures, the area outside the SRV starts to deplete (Fig. 4.19). The significant permeability contrast between SRV and the unstimulated reservoir makes the outer region to appear as a closed boundary. On log-log plot this is identified by overlapping unit-slope $\Delta p/q$ and its derivative. Since the production over this period is mostly from the SRV, the corresponding flow is called SRV flow, as introduced above.

Once the SRV flow is identified, the corresponding data can be used to estimate the SRV area from the slope and intercept of FMB plot ($q/\Delta p$ versus $Q/(c_t \Delta p)$):

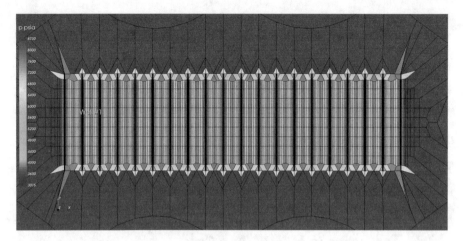

Fig. 4.18 The pressure distribution in the reservoir during transitional flow at 240 h ($t_{MB} = 472.9$ h)

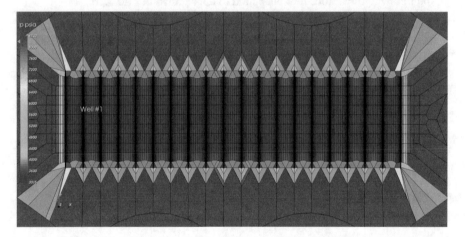

Fig. 4.19 The pressure distribution in the reservoir during SRV flow at 1752 h ($t_{MB} = 9130$ h)

$$\frac{B(1 - S_{wi})}{NB_i} = \frac{\text{slope}}{\text{intercept}} \rightarrow \frac{B}{Ah\phi} = \frac{\text{slope}}{\text{intercept}} \rightarrow A = \frac{B}{h\phi} \frac{\text{intercept}}{\text{slope}} \quad (4.41)$$

The area of the SRV found above equals $2\,x_f\,L_w$ where L_w is the well lateral length.

4.2.3 RTA of MFHWs

The unknown MFHW parameters include: number of fractures (n_f), fracture half-length (x_f), fracture conductivity ($k_f w_f$), and formation permeability (k). The fracture conductivity can be only obtained based on very early-time data. However, this data is generally very noisy and of low quality. The specialized RTA of MFHWs should follow the steps below to obtain n_f, x_f, and k:

1. Identify linear and SRV flow from the log-log plot
2. Start with FMB plot of SRV period's data and obtain x_f considering the known L_w
3. Use square-root time plot of linear flow period's data to obtain $n_f \sqrt{k}$ based on the x_f from step 2
4. Assume k and obtain n_f value
5. Use numerical simulation to match the data by varying n_f and k while $n_f \sqrt{k}$ is fixed
6. Obtain k iteratively

4.2.4 Application to MFHW Base Case

In the following this procedure is applied to data shown in Fig. 4.16.

1. Linear flow is observed on t_{MB} between 10 and 100 h. This corresponds to data between 5.2 and 50 h of actual time. SRV flow is observed between 3000 to 10,000 h of t_{MB} which corresponds to the actual time between 965 and 1840 h.
2. FMB plot ($q/\Delta p$ versus $Q/(c_t \Delta p)$) for the whole production period is shown in Fig. 4.20. Note that $c_t = c_o + c_r = 1.96912 \times 10^{-5}$/psi.

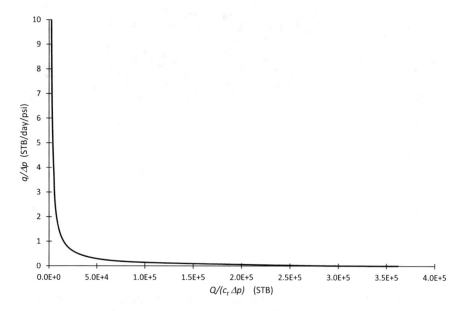

Fig. 4.20 FMB plot for the whole production period

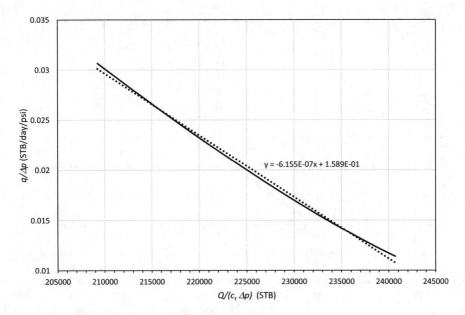

Fig. 4.21 FMB plot during the SRV flow

The FMB plot for the SRV period (965 to 1840 h) is shown in Fig. 4.21. Based on the slope and intercept of the line fitted to this data, we get:

$$A = \frac{B}{h\phi} \frac{\text{intercept}}{\text{slope}} = \frac{2.2}{200 \times 0.05} \frac{0.1589}{6.155 \times 10^{-7}} (5.615) = 318,910 \text{ ft}^2$$

Given that A = 2 Lw x_f, we obtain x_f = 318,910/(2 × 1000) = 159.5 ft.

The actual x_f is 140 ft. Note that SRV flow is not a pure closed boundary flow but a pseudo-closed boundary, hence x_f is overestimated.

3. The square-root time plot over the linear flow period (between 5.2 and 50 h of actual time) is shown in Fig. 4.22. Using the slope, we can get:

$$n_f x_f \sqrt{k} = \frac{31.3B}{\alpha h} \sqrt{\frac{\mu}{\phi c_t}} = \frac{31.3 \times 2.2}{2.9324 \times 200} \sqrt{\frac{0.2}{0.05 \times 19.7e - 6}} = 52.9 \text{ ft}\sqrt{mD}$$

The actual value of $n_f x_f \sqrt{k} = 20 \times 140\sqrt{4 \times 10^{-4}} = 56 \text{ ft}\sqrt{mD}$.

4. Any n_f can be assumed to obtain permeability of SRV to be used in construction of the numerical model. Assuming n_f = 20, and given x_f = 159.5 ft calculated from step 2, we can obtain k = 2.75 × 10^{-4} mD.

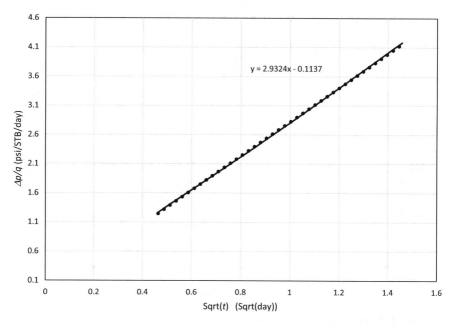

Fig. 4.22 Square-root time plot for the linear flow period

4.3   Complications

In the above, we introduced the rate transient behavior in shale wells considering the following major assumptions:

1. Permeability of the reservoir and the fractures are constant and independent of time. This assumption is questionable especially for the fracture permeability. The permeability of the reservoir and fractures is highly stress/pressure dependent.
2. The fractures are bi-wing with identical width and conductivity over the entire length of the fractures. This assumption is not valid, in practice. The fractures are rarely bi-wing. Also, maximum width of fractures (can be ~1 cm) is at the well and reduces toward the fracture tip. Therefore, the fracture conductivity is variable over the fracture length, in practice.
3. The fractures are completely separate from one another. This assumption is generally invalid in practice since the hydraulically induced fractures may be connected to one another e.g. because of rock heterogeneities or operational variations from one stage to another. More importantly, the hydraulic fractures may be connected to one another via the pre-existing natural fractures.

Accounting for these complications is essential and requires modifications to the analytical and numerical tools introduced herein. The first assumption can be relaxed in the analytical models by including the pressure dependence of the permeability in pseudo-functions. However, assumptions 2 and 3 are difficult to relax analytically. Numerical simulation modifications have been introduced to account for the complex fracture networks the properties of which may vary with time and space.

Also, Darcy equation was used throughout this chapter to model the flow in the fractures and matrix. Darcy flow should be applicable for free flow of hydrocarbons in the fractures (bulk diffusion). However, the following complications make the Darcy flow assumption questionable at pore scale:

1. The produced hydrocarbon in shale may be initially in the form of kerogen. Gas may be produced due to gas desorption from kerogen surface as thermodynamic equilibrium between kerogen and gas phase in the pores changes.
2. Gas diffusion occurs within the kerogen to migrate gas from bulk of kerogen to its surface. Given the very small pores (nano-meter size pore diameters), gas slippage (Klinkenberg effect) can become important at reservoir conditions. The small pores sizes (which is in the same order of magnitude as the hydrocarbon molecules they confine) affect the PVT behavior of the hydrocarbons. Primarily, they affect the bubble/dew point of the hydrocarbons and the phase diagram.

Accounting for intra-pore and inter-pore gas diffusion combined with desorption requires additional modelling considerations.

Addressing the abovementioned complications is the subject of ongoing research efforts and thus, beyond the scope of this book. The main point is that the RTA methodology presented above may not be accurately applicable and should be used with caution. Nevertheless, it should give a good starting point.

Appendix 1: Pressure-Rate Relationship for Boundary-Dominated Flow Under Constant Rate

For constant rate production, the average pressure is related to the well flowing pressure by:

$$\bar{p} = p_{wf} + \frac{q\mu B}{2\pi kh} \left[\frac{1}{2} \ln \left(\frac{4A}{e^\gamma C_A r_w^2} \right) + s \right] \tag{4.42}$$

where B should be evaluated at $\frac{\bar{p}+p_{wf}}{2}$. Also,

$$\bar{p} - p_i = \frac{-qB(1-S_{wi})}{NB_i c_t} t \tag{4.43}$$

In this equation, B should be evaluated at $\frac{\bar{p}+p_i}{2}$. For convenience, we assume that the formation volume factor is relatively constant.

The latter equation can be derived using the definition of total compressibility:

$$c_t = \frac{1}{V_p} \frac{dV_p}{d\bar{p}} \tag{4.44}$$

According to this equation, withdrawal of dV_p volume of oil from the pore space of a single-phase oil reservoir with pore volume of V_p and total compressibility c_t results in the drop in the average pressure of $d\bar{p}$. Note that: $V_p = Ah\phi = NB_i$ where A, h, and ϕ are the reservoir area, thickness, and initial porosity respectively. N and B_i are the oil initial in place (assuming no water in the system) and initial oil formation volume factor respectively. Also, $dV_p = -QB$ where Q is the cumulative production at standard conditions and B is evaluated at the average pressure i.e. $\frac{\bar{p}+p_i}{2}$. Consequently, $\frac{dV_p}{dt} = -qB$ where q is production rate at standard conditions. Replacement in the compressibility while dividing the numerator and denominator by dt gives:

$$c_t = \frac{1}{Ah\phi} \frac{-qB}{d\bar{p}/dt} \tag{4.45}$$

Therefore,

$$\frac{d\bar{p}}{dt} = \frac{-qB}{Ah\phi c_t} = \frac{-qB}{\pi r_e^2 h\phi c_t} \tag{4.46}$$

Integration results in the second equation shown in this appendix. Combining the first two equations above gives:

$$p_i = p_{wf} + \frac{q\mu B}{2\pi kh} \left[\frac{1}{2} \ln\left(\frac{4A}{e^\gamma C_A r_w^2} \right) + s \right] + \frac{qB(1-S_{wi})}{NB_i c_t} t \tag{4.47}$$

Or

$$\frac{p_i - p_{wf}}{q} = \frac{\mu B}{2\pi kh} \left[\frac{1}{2} \ln\left(\frac{4A}{e^\gamma C_A r_w^2} \right) + s \right] + \frac{B(1-S_{wi})}{NB_i c_t} t \tag{4.48}$$

Note: In this equation, B in the first term on the RHS is the average formation volume factor evaluated at $\frac{\bar{p}+p_{wf}}{2}$. B in the second term on the RHS is the average formation volume factor evaluated at $\frac{\bar{p}+p_i}{2}$. For convenience, we assume the formation volume factor is relatively constant and therefore, no distinction is made between these two formation volume factors.

Appendix 2: Rate-Pressure Relationship for Linear Flow Under Constant Rate Production

Consider the same linear from governing equation. However, let's subject it to constant rate boundary condition:

$$\frac{\partial^2 \Delta p}{\partial x^2} = \frac{1}{\eta} \frac{\partial \Delta p}{\partial t} \tag{4.49}$$

Subject to:

$$\Delta p(x, t = 0) = 0 \tag{4.50}$$

$$\Delta p(x \to \infty, t) = 0 \tag{4.51}$$

$$\frac{d\Delta p(x = 0, t)}{dx} = -\frac{q\mu B}{kA} \tag{4.52}$$

Laplace transform of the above gives:

$$\frac{d^2 \Delta \bar{p}}{\partial x^2} - \frac{S}{\eta} \Delta \bar{p} = 0 \tag{4.53}$$

Subject to:

$$\Delta \bar{p}(x \to \infty) = 0 \tag{4.54}$$

$$\frac{d\Delta \bar{p}(x = 0, t)}{dx} = -\frac{q\mu B}{kA} \frac{1}{S} \tag{4.55}$$

The general solution is:

$$\Delta \bar{p} = C_3 e^{-x\sqrt{S/\eta}} + C_4 e^{x\sqrt{S/\eta}} \tag{4.56}$$

The first BC results in C4 = 0. Therefore,

$$\frac{d\Delta \bar{p}}{dx} = -C_3 \sqrt{\frac{S}{\eta}} e^{-x\sqrt{S/\eta}} \tag{4.57}$$

Applying the second BC gives:

$$-C_3 \sqrt{\frac{S}{\eta}} = -\frac{q\mu B}{kA} \frac{1}{S} \tag{4.58}$$

Therefore:

$$C_3 = \frac{q\mu B}{kA} \frac{1}{S} \sqrt{\frac{\eta}{S}} \tag{4.59}$$

And the solution is:

$$\Delta \bar{p} = \frac{q\mu B}{kA} \frac{1}{S} \sqrt{\frac{\eta}{S}} e^{-x\sqrt{S/\eta}} \tag{4.60}$$

Inversion gives:

$$\Delta p = \frac{q\mu B}{kA} \left(\frac{\frac{2e^{-\frac{x^2}{4\eta t}}\sqrt{t}}{\sqrt{\pi}} - \frac{x}{\sqrt{\eta}} + \frac{xErf\left[\frac{x}{2\sqrt{\eta t}}\right]}{\sqrt{\eta}}}{\sqrt{\frac{1}{\eta}}} \right) \tag{4.61}$$

Setting x = 0 to obtain the pressure at the sandface gives:

$$p_i - p_{wf} = \frac{2}{\sqrt{\pi}} \frac{q\mu B}{kA} \sqrt{\eta t} \tag{4.62}$$

Therefore,

$$\frac{p_i - p_{wf}}{q} = \frac{2}{\sqrt{\pi}} \frac{\mu B}{kA} \sqrt{\eta t} \tag{4.63}$$

Therefore,

$$\frac{p_i - p_{wf}}{q} = \frac{1}{2\sqrt{\pi}} \frac{B}{x_f h} \sqrt{\frac{\mu}{\phi k c_t}} \sqrt{t} \tag{4.64}$$

In field units with time in hours, this equation is:

$$\frac{p_i - p_{wf}}{q} = 4.06 \frac{B}{x_f h} \sqrt{\frac{\mu}{\phi k c_t}} \sqrt{t} \tag{4.65}$$

And with time in days, the equation becomes:

$$\frac{p_i - p_{wf}}{q} = 19.89 \frac{B}{x_f h} \sqrt{\frac{\mu}{\phi k c_t}} \sqrt{t} \tag{4.66}$$

Appendix 3: Rate-Pressure Relationship for Linear Flow Under Constant Well Flowing Pressure

The governing equation for linear flow in terms of $\Delta p = p_i - p_{wf}$ is:

$$\frac{\partial^2 \Delta p}{\partial x^2} = \frac{1}{\eta} \frac{\partial \Delta p}{\partial t}$$

(4.67)

Subject to:

$$\Delta p(x, t=0) = 0$$

(4.68)

$$\Delta p(x \rightarrow \infty, t) = 0$$

(4.69)

$$\Delta p(x=0, t) = p_i - p_{wf} = c$$

(4.70)

Laplace transform of the above gives:

$$\frac{d^2 \Delta \bar{p}}{\partial x^2} - \frac{S}{\eta} \Delta \bar{p} = 0$$

(4.71)

Subject to:

$$\Delta \bar{p}(x \rightarrow \infty) = 0$$

(4.72)

$$\Delta \bar{p}(x=0) = \frac{c}{S}$$

(4.73)

Therefore:

$$\Delta \bar{p} = C_1 e^{-x\sqrt{S/\eta}} + C_2 e^{x\sqrt{S/\eta}}$$

(4.74)

Using the first boundary condition, $C_2 = 0$. Therefore:

$$C_1 = \frac{c}{S}$$

(4.75)

In summary the solution is:

$$\Delta \bar{p} = (p_i - p_{wf}) \frac{e^{-x\sqrt{S/\eta}}}{S}$$

(4.76)

Therefore, the well rate is given by:

$$\bar{q} = -\frac{kA}{\mu B}\frac{d\Delta\bar{p}}{dx} = \frac{kA}{\mu B}\frac{p_i - p_{wf}}{s}\sqrt{S/\eta}\,e^{-x\sqrt{S/\eta}} \tag{4.77}$$

The following Laplace inversion is available:

$$L^{-1}\left\{\frac{\sqrt{S/\eta}\,e^{-x\sqrt{S/\eta}}}{s}\right\} = e^{-\frac{x^2}{4\eta t}}\sqrt{\frac{1}{\pi\eta t}} \tag{4.78}$$

Therefore,

$$q = \frac{kA}{\mu B}\left(p_i - p_{wf}\right)e^{-\frac{x^2}{4\eta t}}\sqrt{\frac{1}{\pi\eta t}} \tag{4.79}$$

Rearrangement gives:

$$\frac{\left(p_i - p_{wf}\right)}{q} = \frac{\mu B}{kA}\,e^{\frac{x^2}{4\eta t}}\sqrt{\pi\eta t} \tag{4.80}$$

Setting $x = 0$ to obtain rate at the sandface gives:

$$\frac{\left(p_i - p_{wf}\right)}{q} = \frac{\mu B}{kA}\sqrt{\pi\eta t} \tag{4.81}$$

Replacement of $A = 4\,x_f\,h$ gives:

$$\frac{\left(p_i - p_{wf}\right)}{q} = \frac{\sqrt{\pi}}{4}\frac{B}{hx_f}\sqrt{\frac{\mu}{\phi c_t k}}\sqrt{t} \tag{4.82}$$

In field units with time in hours, this equation is:

$$\frac{\left(p_i - p_{wf}\right)}{q} = 6.39\frac{B}{hx_f}\sqrt{\frac{\mu}{\phi c_t k}}\sqrt{t} \tag{4.83}$$

And with time in days, the equation becomes:

$$\frac{\left(p_i - p_{wf}\right)}{q} = 31.3\frac{B}{hx_f}\sqrt{\frac{\mu}{\phi c_t k}}\sqrt{t} \tag{4.84}$$

References

Blasingame T, McCray T, Lee W (1991) Decline curve analysis for variable pressure drop/variable flowrate systems. SPE Gas Technology Symposium. OnePetro

Burton A (2016) An overview of multistage completion systems for hydraulic fracturing, society of petroleum engineers. https://webevents.spe.org/products/an-overview-of-multistage-completion-systems-for-hydraulic-fracturing#tab-product_tab_overview

Charlez P (2019) Total Unconventional Seminar, Baton Rouge, LA

Chen H, Meng X, Niu F, Tang Y, Yin C, Wu F (2018) Microseismic monitoring of stimulating shale gas reservoir in SW China: 2. Spatial clustering controlled by the Preexisting faults and fractures. J Geophys Res Solid Earth 123:1659–1672

Cheng Y, Lee WJ, McVay DA (2008) Improving reserves estimates from decline-curve analysis of tight and multilayer gas wells. SPE Reserv Eval Eng 11:912–920

Ecoflight (2006). https://ecoflight.zenfolio.com/p648196342/h32638d19#h32638d19

Economides MJ, Hill AD, Ehlig-Economides C, Zhu D (2013) Petroleum production systems. Pearson Education

Elton DC (2018) Stretched exponential relaxation. arXiv preprint arXiv:1808.00881

Elwaziry Y (2020) Hydraulic fracturing operations. https://www.youtube.com/watch?v=ot6F42szK9c&ab_channel=PioPetro

Gale JFW, Laubach SE, Olson JE, Eichhubl P, Fall A (2014) Natural fractures in shale: a review and new observations. AAPG Bull 98:2165–2216

Hartog AH (2017) An introduction to distributed optical fibre sensors. CRC Press, FL

Houze O, Viturat D, Fjaere O (2020) Dynamic Data Analysis (DDA)

Houze O, Viturat D, Fjaere O (2022) Dynamic Data Analysis (DDA)

Hveding F, Guraini W, Mahue V, Hafezi S (2020) Production flow comparison between distributed Fiber-optic sensing and conventional PLT in a cased hole horizontal wellbore with ICD. Abu Dhabi International Petroleum Exhibition & Conference

Ilk D, Rushing JA, Perego AD, Blasingame TA (2008) Exponential vs. hyperbolic decline in tight gas sands: understanding the origin and implications for reserve estimates using Arps' decline curves

Kappa-Rubis (2020) Kappa Version 2020 user's guide. Kappa Eng, Houston, TX

Liu Y, Jin G, Wu K, Moridis G (2022) Quantitative hydraulic-fracture-geometry characterization with low-frequency distributed-acoustic-sensing strain data: fracture-height sensitivity and field applications. Spe Prod Oper 37:159–168

Mao Y, Godefroy C, Gysen M (2021) Field applications of flow-Back profiling using distributed temperature data. SPE/AAPG/SEG Asia Pacific unconventional resources technology conference

Natareno N, Thelen M, Charbonneau J, Sahdev N, Cook P (2019) Continuous use of Fiber optics-enabled coiled tubing used to accelerate the optimization of completions aimed at improved recovery and reduced cost of development. SPE hydraulic fracturing technology conference and exhibition

Pakhotina I, Sakaida S, Zhu D, Hill AD (2020) Diagnosing multistage fracture treatments with distributed Fiber-optic sensors. Spe Prod Oper 35:0852–0864

Passey QR, Bohacs KM, Esch WL, Klimentidis R, Sinha S (2010) From oil-prone source rock to gas-producing shale reservoir – geologic and petrophysical characterization of unconventional shale-gas reservoirs. International oil and gas conference and exhibition in China

Raterman KT, Farrell HE, Mora OS, Janssen AL, Gomez GA, Busetti S, McEwen J, Davidson M, Friehauf K, Rutherford J, Reid R, Jin G, Roy B, Warren M (2017) Sampling a stimulated rock volume: an eagle ford example. SPE/AAPG/SEG unconventional resources technology conference

Rushing JA, Perego AD, Sullivan R, Blasingame TA (2007) Estimating reserves in tight gas sands at HP/HT reservoir conditions: use and misuse of an Arps decline curve methodology. SPE annual technical conference and exhibition. OnePetro

Shahri M, Tucker A, Rice C, Lathrop Z, Ratcliff D, McClure M, Fowler G (2021). High fidelity fibre-optic observations and resultant fracture modeling in support of planarity. SPE hydraulic fracturing technology conference and exhibition. OnePetro

Sookprasong PA, Hurt RS, Gill CC, LaFollette RF (2014) Fiber optic DAS and DTS in multicluster, multistage horizontal well fracturing: interpreting hydraulic fracture initiation and propagation through diagnostics. SPE annual technical conference and exhibition

Spivey J, Frantz J, Williamson J, Sawyer W (2001) Applications of the transient hyperbolic exponent. SPE Rocky Mountain petroleum technology conference. OnePetro

US-EIA (2014) Updates to the EIA eagle ford play maps. U.S. Dept energy, Washington, DC

US-EIA (2020) U.S. crude oil production grew 11% in 2019, surpassing 12 million barrels per day. U.S. Energy Information Administration. https://www.eia.gov/todayinenergy/detail.php?id=43015#

US-EIA (2022) Annual energy outlook 2022 with projections to 2050. Energy Information Administration, United States Department of Energy, Washington DC

Valko PP (2009) Assigning value to stimulation in the Barnett shale: a simultaneous analysis of 7000 plus production hystories and well completion records. SPE hydraulic fracturing technology conference. OnePetro

Valkó PP, Lee WJ (2010) A better way to forecast production from unconventional gas wells. SPE annual technical conference and exhibition. OnePetro

Weber M, Weatherly D, Mahue V, Hull RA, Trujillo K, Bohn R, Jimenez E (2021) Optimizing completion designs for the East Texas Haynesville utilizing production flow allocations from lower-cost Fiber optic sensing DAS/DTS systems. SPE/AAPG/SEG unconventional resources technology conference

Weijers L, Wright C, Mayerhofer M, Pearson M, Griffin L, Weddle P (2019) Trends in the north American Frac industry: invention through the shale revolution. SPE hydraulic fracturing technology conference and exhibition

Wickstrom L, Erenpreiss M, Riley R, Perry C, Martin D (2012) Geology and activity update of the Ohio Utica-point pleasant play

Wright JDW (2015) Oil and gas property evaluation. Thompson-Wright, LLC

Zborowski M (2019) Exploring the innovative evolution of hydraulic fracturing. J Pet Technol 71:39–41

Zhang Y, Lv D, Wang Y, Liu H, Song G, Gao J (2020) Geological characteristics and abnormal pore pressure prediction in shale oil formations of the Dongying depression, China. Energy Sci Eng 8:1962–1979

Printed in the United States
by Baker & Taylor Publisher Services